Contents

Unit B1: Influences on Life

Unit B2: The Components of Life

Unit B3: Using Biology

Answers (found at the centre of the book)

Questions labelled with an asterisk (*) are ones where the quality of your written communication will be assessed – you should take particular care with your spelling, punctuation and grammar, as well as the clarity of expression, on these questions.

1. State the **five** kingdoms used to classify all living organisms. (5)

(a) ....

(b) ....

(c) ....

(d) ....

(e) ....

*2. Describe the characteristics of the organisms within each of these five kingdoms. (6)

....

....

....

....

....

....

....

....

3. Suggest why grouping animals by colour would be an unsatisfactory method of classification. (1)

....

....

4. You have found a new vertebrate in a tropical rainforest and want to begin classifying it. Suggest **two** immediate questions you would ask to begin this process. (2)

....

....

....

5. When classifying an organism, the groups get smaller and smaller. State what you would expect to see occurring between the organisms within the groups. (2)

....

Nick Dixon

Exam Practice
Workbook

Edexcel
GCSE Biology

Contents

6. **(a) (i)** State **one** animal that is difficult to classify accurately. (1)

(ii) Explain why this animal is difficult to classify accurately. (1)

(b) Explain why scientists have found it difficult to classify viruses. (2)

7. The animal kingdom can be divided into vertebrates and invertebrates. State how animals in these groups are different from each other. (1)

*8. Describe the characteristics of organisms within the **five** vertebrate classes. (6)

9. Describe the characteristics of organisms within the invertebrate phyla and classes listed below. (5)

Annelids ....

....

Molluscs ....

....

Crustaceans ....

....

Arachnids ....

....

Insects ....

....

10. Identification keys are used to name living organisms. Pam was looking at some arthropods in her local zoo. State the **five** arthropods she looked at, using the key below. (5)

Question 1: Does it have claws or pincers? If yes, go to question 2. If not, move onto question 3.

Question 2: Does it have antennae? If yes, it is a lobster. If not, it is a scorpion.

Question 3. Does it have wings? If yes, it is a bee. If not, move on to question 4.

Question 4. Does it have four pairs of legs? If yes, it is a spider. If not, it is a woodlouse.

A ....

B ....

C ....

D ....

E ....

11. What is the name of the phylum that contains vertebrates? (1)

A ◯ Mollusca   B ◯ Chordata

C ◯ Nematoda   D ◯ Porifera

12. (a) Cacti and camels have evolved to live in hot environments. Explain **two** ways in which each organism is adapted to survive. (4)

_____

_____

_____

_____

_____

_____

_____

(b) Other organisms have evolved to survive in extreme habitats. State **two** examples of these habitats. (2)

_____

_____

_____

13. Explain **two** ways in which some predators are adapted to hunt and some prey are adapted to escape. (4)

_____

_____

_____

_____

_____

14. (a) State the definition of the term **variation**. (1)

_____

_____

(b) State **two** examples of variation in humans. (2)

....

....

15. State the **two** factors that can cause variation. Suggest an example for each factor.

(a) Factor: .... Example: .... (2)

(b) Factor: .... Example: .... (2)

16. Below is a list of features that can be used to describe someone. State whether each feature is genetic, environmental or a combination of both. (3)

(a) 1.74m tall .... **(b)** Hazel eyes ....

(c) Scar on lower lip ....

17. Describe what effect genetic and environmental factors can have upon sporting ability. (3)

....

....

....

....

....

....

18. **(a)** Variation can be divided into two categories: discontinuous and continuous. State **two** examples in humans for each type given below. (4)

Discontinuous ....

Continuous ....

(b) Describe how you would show results from surveys into discontinuous and continuous variation. (2)

....

....

....

(c) The table below shows the percentage of the UK population with different blood groups.

Blood Group	Percentage of UK Population
O	44
A	42
B	10
AB	4

(i) Draw a graph to show this. (3)

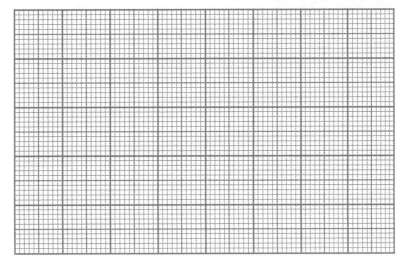

(ii) Describe these results. (3)

....

....

....

(d) Draw a line graph to show the distribution you would expect to see in the number of students with different heights in a year group at secondary school. (1)

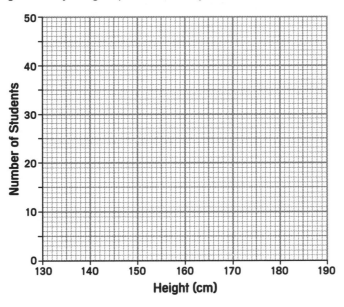

19. **(a)** Describe **three** ways in which scientists support, disprove or improve scientific evidence. (3)

....

....

....

(b) Evidence for Darwin's theory of evolution comes from: (1)

A ☐ the extinction of the dodo

B ☐ increasing rates of diabetes

C ☐ the emergence of antibiotic-resistant bacteria

D ☐ longer average life spans of UK citizens.

*20. Charles Darwin (1809–1882) was a scientist who proposed the theory of evolution by natural selection. Describe the **six** steps in natural selection. (The six steps must be given in the correct order.) (6)

....

....

....

....

....

....

....

....

....

21. Evidence for evolution can be seen in the fossil record.

(a) State what **fossils** are. (1)

....

....

(b) Suggest **three** reasons why the fossil record is incomplete.   (3)

22.   Describe how humans and apes evolved from a common ancestor.   (2)

23.   Use numbers **1** to **4** to put the following in order of their size from the largest to the smallest.   (4)

Gene ....   Chromosome ....

Nucleus ....   Cell ....

24.   **(a)** State the definition of the term **genes**.   (1)

(b) State what **chromosomes** are.   (1)

(c) State how many pairs of chromosomes there are in a human body cell.   (1)

(d) Describe what our genes control.   (1)

25.   **(a)** State the definition of the term **alleles**.   (1)

(b) State the name for an allele that controls the development of a characteristic only when it is present on both chromosomes.   (1)

26. Cystic fibrosis is a genetic disorder, caused by a recessive gene inherited from both parents. (2)

(a) Describe the symptoms of this condition.

Sickle cell disease is also a genetic disorder, caused by recessive alleles. (2)

(b) Describe the symptoms of this condition.

27. You have inherited an allele from each of your parents for eye colour. State the term that scientists use for each of the **three** combinations of genotypes that you might have inherited. (3)

(a) Two alleles for brown eyes

(b) Two alleles for blue eyes

(c) One allele for brown eyes and one for blue eyes

28. (a) State the definition of the term **genotypes**. (1)

(b) State the definition of the term **phenotypes**. (1)

29. State whether the following genotypes are **homozygous dominant, homozygous recessive** or **heterozygous**. (3)

(a) Bb

(b) BB

(c) bb

30. The gene that controls earlobe type has two alleles. The allele for attached lobes is recessive (e) and the allele for unattached lobes is dominant (E). (3)

State the possible combinations of alleles and, for each one, name the phenotype.

31. Complete these two different crosses between a brown-eyed parent and a blue-eyed parent (brown eyes are dominant). (4)

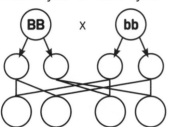 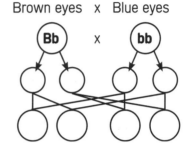

32. Explain how two parents with brown eyes could produce a child with blue eyes. Draw a diagram to help you. (4)

33. **(a)** Complete the following Punnett square diagrams to show how attached and unattached earlobes might be inherited (E is the dominant allele for unattached lobes). (3)

(i)

	E	e
E		
e		

(ii)

	e	e
E		
e		

(iii)

	e	e
e		
e		

(b) State the percentage chances of producing a child with unattached earlobes from each of the crosses in **(a)**. (3)

(i) ....

(ii) ....

(iii) ....

34. Brown eyes are dominant over blue eyes. A homozygous dominant male mates with a heterozygous female. Draw a genetic diagram to show the probable genotypic ratios of the offspring. Use genetic terminology to describe them. (4)

35. **(a)** Unattached ear lobes are dominant over attached lobes. A heterozygous male mates with a heterozygous female. Draw a genetic diagram to show the probable genotypic ratios of the offspring. Use genetic terminology to describe them. (4)

(b) A homozygous recessive individual mates with their homozygous recessive partner. Their offspring mates with a heterozygous partner. Draw a genetic diagram to show the probable genotypic ratios of their offspring. Use genetic terminology to describe the offspring. (5)

36. Unattached ear lobes are dominant. A homozygous dominant individual mates with their homozygous dominant partner. Their child mates with a heterozygous partner. Draw a genetic diagram to show the probable genotypic ratios of their offspring. Use genetic terminology to describe the offspring. (5)

(Total: .... / 147)

Higher Tier

37. **(a)** Describe the significance of hybrid ducks in terms of classification. (3)

continued...

(b) (i) State the definition of the term **ring species**. (1)

(ii) Describe the significance of ring species in terms of reproduction. (1)

(iii) State an example of a ring species. (1)

38. State what the binomial system of classification gives each organism. (1)

39. Describe why classification is important. (1)

40. Explain how sickle cell disease is inherited. Include an explanation of the term **carriers** in your answer. (2)

41. **(a)** State the definition of the term **speciation**. (1)

(b) Describe how speciation might occur. (1)

(c) State an example of an organism that has undergone this process. (1)

(Total: _____ / 13)

Questions labelled with an asterisk (*) are ones where the quality of your written communication will be assessed – you should take particular care with your spelling, punctuation and grammar, as well as the clarity of expression, on these questions.

1. **(a)** State the definition of the term **homeostasis**. (1)

    _____

    _____

    (b) State **two** common examples of homeostasis. (2)

    _____

    _____

    (c) (i) Explain why homeostasis is important. (1)

    _____

    (ii) Describe what might happen if homeostasis is not maintained. (1)

    _____

2. Describe what happens to our urine when we do not drink enough water. State the colour
    it becomes in your answer. (2)

    _____

    _____

3. Describe the role of the skin in regulating our temperature when we become too cold. (4)

    _____

    _____

    _____

    _____

4. The temperature of the human body has to be maintained.

    (a) State the normal body temperature in humans. (1)

    _____

    (b) Explain why this temperature is maintained. (1)

    _____

    (c) State the part of the brain that monitors temperature. (1)

    _____

5.   **(a)** Explain why it is an advantage for an organism to possess a nervous system.   (1)

....

(b) Label the **three** parts of the nervous system shown on the diagram below.   (3)

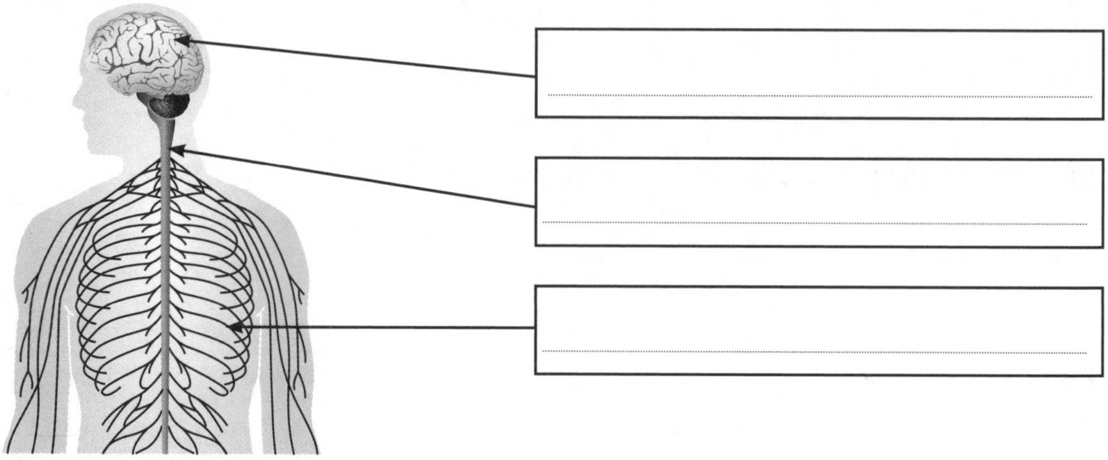

6.   State the sense associated with each organ in the table below. The first one has been done for you.   (4)

Organ	Sense
(a) Ears	Sound
(b) Eyes	
(c) Skin	
(d) Tongue	
(e) Nose	

7.   Draw a straight line to correctly match each different part of the brain to its function.   (4)

Part of the Brain

| Cerebellum |
| Medulla |
| Frontal lobe |
| Cerebral hemisphere |

Function

| Coordinates movement and balance |
| Responsible for numerical computation, language and emotions |
| Controls automatic actions like heart beat and breathing |
| Controls higher mental functions like choice and memory |

8. Draw and label diagrams to show the **three** different types of nerve cell. Use arrows to show the direction in which an electrical impulse travels along each one. (3)

9. The diagram below shows a motor neurone.

Explain how a motor neurone is specially adapted for its purpose. (3)

....

....

....

10. The diagram below shows the junction between two neurones.

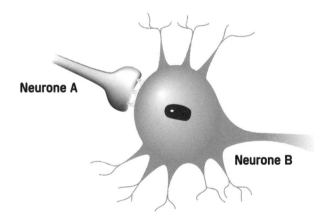

Neurone A

Neurone B

(a) Draw an arrow on the diagram to show the direction in which nerve impulses travel across this junction. (1)

(b) State the name given to the gap between two neurones. (1)

....

(c) Describe how nerve impulses travel across this gap. (4)

11. Number the following parts of the nervous system **3** to **6** to show the pathway that a nerve impulse follows through the body in response to a stimulus. (4)

(a) Effector (e.g. muscle) `7`

(b) Synapse between sensory neurone and relay neurone ` `

(c) Receptor (in sense organ) `1`

(d) Motor neurone ` `

(e) Relay neurone ` `

(f) Synapse between relay neurone and motor neurone ` `

(g) Sensory neurone `2`

12. **(a)** Describe what is meant by the terms voluntary and involuntary responses. Give an example of each in your answer. (4)

(b) Describe **two** experiments in which you investigate involuntary responses in humans. (2)

13. In the kitchen, Richard reaches out his hand to take a saucepan from the hob. 'Ow, it's hot!' he says, pulling his hand away.

(a) Richard's reaction was a reflex response. Explain why reflex responses like this are important. (1)

....

(b) The diagram below shows the reflex arc involved in this action. Name the parts labelled **A**, **B**, **C** and **D**. (4)

A ....

B ....

C ....

D ....

(c) Describe the pathway of this reflex arc. (4)

....

....

....

....

*14. There are many receptors on the surface of the skin. Describe an experiment in which you investigate how far apart touch receptors are on different parts of your skin. State an area which is likely to have touch receptors close together in your answer. (6)

....

....

....

....

....

15. In addition to the nervous system, the human body uses chemical messages to help control its functions.

(a) State what these substances are called. _____ (1)

(b) State where they are produced. _____ (1)

(c) Describe how they travel around the body. _____ (1)

(d) State **three** different chemical messages. For each example, name the part of the body where it is produced. (3)

(i) _____

(ii) _____

(iii) _____

16. **(a)** If there is too much glucose in our blood, which hormone does the pancreas produce? (1)

A ◯ Oestrogen

B ◯ Glucagon

C ◯ Progesterone

D ◯ Insulin

(b) State what the medical condition **diabetes** is. (1)

_____

_____

17. Describe how blood glucose levels are lowered in people who do not have diabetes. (3)

_____

_____

_____

_____

18. **(a)** State **three** symptoms of diabetes. (3)

_____

_____

_____

(b) Describe the difference between the two types of diabetes, by completing the
following sentences. (2)

   (i) Type 1 diabetes sufferers are: ....

   (ii) Type 2 diabetes sufferers are: ....

(c) Suggest **three** things diabetics should do on a daily basis, besides injecting insulin,
to reduce the effects of their condition. (3)

....

....

....

19. The graph shows the blood sugar level for a person with diabetes who did not take insulin for a
12-hour period.

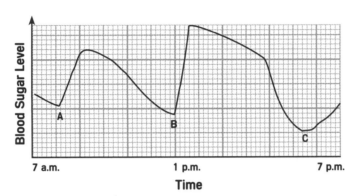

(a) Explain how we can tell from the graph that this person has diabetes. (1)

....

(b) Explain why the blood sugar level rises rapidly at points A and B. (1)

....

(c) Describe what would happen at points A and B if the person did not have diabetes. (1)

....

(d) Suggest why the person needed to eat a chocolate bar at point C. (1)

....

20. (a) State the definition of the terms **positively phototropic** and **positively geotropic**. (2)

....

....

....

*(b) Describe an experiment in which you could investigate phototropism in plants. Explain the results you would expect to find. Mention a 'control' in your answer. (6)

(c) The diagram below shows a shoot tip growing towards the light. Draw and label an arrow to show the direction of light. Draw approximately sized plant cells on either side of the shoot to show the action of auxins. (3)

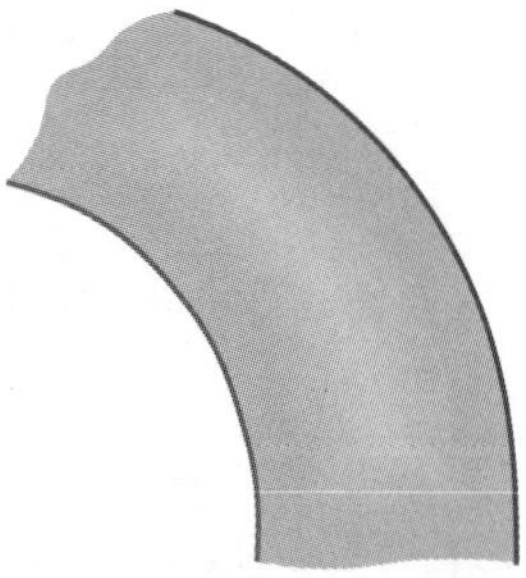

21. There are **two** main types of plant hormones. State their names and describe what they do. (2)

(Total: _____ **/ 99)**

22. The two diagrams show a blood vessel and a sweat gland in the skin.

(a) Which diagram shows skin in hot conditions? (1)

....

(b) Describe what happens when the body's temperature gets too low. (4)

....

....

....

....

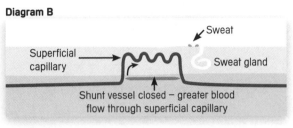

Diagram A

Sweating stopped

Superficial capillary → Sweat gland

Shunt vessel open – reduced blood flow through superficial capillary

Diagram B

Sweat

Superficial capillary → Sweat gland

Shunt vessel closed – greater blood flow through superficial capillary

(c) Describe what happens when the body's temperature gets too high. (4)

....

....

....

....

23. Describe what happens when blood glucose levels are too low. (3)

....

....

....

....

24. Describe **three** commercial uses of plant hormones. (3)

....

....

....

(Total: .... / 15)

Questions labelled with an asterisk (*) are ones where the quality of your written communication will be assessed – you should take particular care with your spelling, punctuation and grammar, as well as the clarity of expression, on these questions.

1. **(a)** Describe how caffeine affects the nervous system. (1)

    ....

    (b) State **two** effects caffeine could have on behaviour. (2)

    ....

    ....

2. Explain why paracetamol stops you from feeling pain. (1)

    ....

3. Barbiturates are sedatives. However, in recent years, doctors have started to use them much less. Explain why this is. (1)

    ....

    ....

4. In addition to the effects of the drug on the physical and mental health of addicts, there are other risks associated with heroin use. Suggest **one**. (1)

    ....

5. State **three** things that cannabinoids and opiates have in common.

    (a) .... (1)

    (b) .... (1)

    (c) .... (1)

6. **(a)** State the substance in tobacco that is addictive. .... (1)

    (b) State the substance in tobacco that can cause cancer. .... (1)

    (c) State **three** other diseases caused by smoking. (3)

    (i) .... **(ii)** .... **(iii)** ....

7. **(a)** Describe **two** effects of smoking on the respiratory system and explain their consequences. (4)

    ....

    ....

    ....

(b) Describe **two** effects of smoking on the circulatory system and explain their consequences. (4)

8. **(a)** Describe what effects moderate alcohol consumption has on the body. (2)

(b) State **two** long-term health problems associated with excess alcohol consumption. (2)

9. **(a)** Describe what solvents are. (2)

(b) State **two** effects that solvent use has on the body. (2)

*10. Describe how you could investigate whether caffeine affects reaction times. Remember to include how to make the experiment reliable in your answer. (6)

11. Describe what a transplant is. Give **two** examples of organs or tissues that can be transplanted in your answer. (3)

_____

_____

_____

12. Some people have strong feelings about organ donation. Explain why this is a controversial issue. (3)

_____

_____

_____

_____

_____

13. (a) State the definition of the term **pathogen**. (1)

_____

(b) State **two** types of pathogen. For each one, give an example of a disease they transmit. (4)

_____

_____

_____

_____

14. Pathogens can be transmitted by direct contact and indirect contact.

(a) State which of these methods of transmission is the most common. (1)

_____

(b) State **two** different methods of transmission. For each one, give an example of a relevant infection. (4)

_____

_____

_____

_____

(c) State the definition of the term **vector**. Give an example in your answer. (2)

_____

_____

15. Describe the functions of each of the animal defences listed below. State whether each defence is a physical or chemical barrier. (3)

Skin: .... barrier

Cilia: .... barrier

Lysozymes: .... barrier

16. Aspirin is used to relieve minor pain. It originated from willow plants. State **one** other drug that originated from a plant, and the plant it came from. (2)

....

....

17. **(a)** State the **two** types of antibiotic. State the microorganisms they kill in your answer. (2)

....

....

....

(b) State the type of infection that antibiotics do not treat. (1)

....

18. **(a)** Describe what antiseptics are. (1)

....

....

*(b) Describe an experiment you could complete to investigate the effects of different antiseptics (or antibiotics). Mention a 'control' in your answer. (6)

....

....

....

....

....

....

....

19. State the definition of the term **interdependence**. (1)

....

....

20. **(a)** What does a food chain show? (1)

....

(b) State what the arrows in a food chain indicate. (1)

....

(c) Describe what a pyramid of number is. (1)

....

....

(d) Draw a pyramid of number for the following food chain in the space below. (4)

Rose Bush ⟶ Greenfly ⟶ Blue Tit ⟶ Hawk

(e) Draw a pyramid of biomass for the same food chain in the space below. (4)

(f) Describe what a pyramid of biomass is. (1)

....

....

(g) State the **two** reasons why a pyramid of number might be different in shape from a pyramid of biomass. (2)

....

....

....

21. The diagram below shows how 1000J of energy flows through a food chain.

(a) State the process by which the grass captures energy from the Sun. (1)

....

(b) State a life process in the rabbits that results in the transfer of heat to the surroundings. (1)

....

(c) Calculate the percentage of the original energy from the Sun that the fox receives.

Energy received = ... % (1)

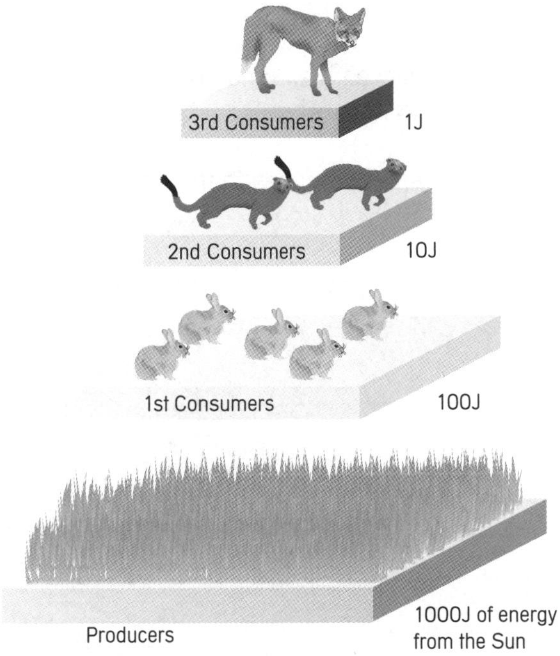

3rd Consumers 1J

2nd Consumers 10J

1st Consumers 100J

1000J of energy from the Sun

Producers

(d) Describe the ways in which energy is lost at each stage in the food chain. (2)

....

....

....

22. Neil's cat seems to be losing weight. He takes it to the vet who tells him that it has a tapeworm.

(a) State the word that describes organisms, such as tapeworms, that benefit from their hosts. (1)

....

(b) State **two** other examples of organisms like this that Neil's cat could be suffering from. (2)

....

23. The population of the world is increasing.

(a) Show this growth on the graph below. (1)

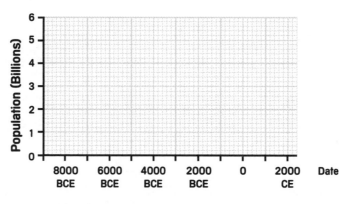

(b) State **two** problems that may be caused by this increase. (2)

_____

24. Peter leaves his home on the Isle of Wight to visit Mexico City. He notices a big difference in the amount of air pollution in the two places.

(a) State **two** gases that cause air pollution. State where they originate from in your answer. (4)

_____

_____

_____

_____

(b) On his way to the airport, he notices a polluted river running into a lake. State **two** chemicals that cause water pollution. State where they originate from in your answer. (4)

_____

_____

_____

_____

(c) Peter stops and looks at the river. The presence or absence of some organisms will tell him how polluted the water is.

(i) State what this type of organism is called. _____ (1)

(ii) Give **two** examples of this type of organism. (2)

_____

*25. Describe an experiment in which you investigate the effects of pollutants on plant growth. Mention a 'control' in your answer.   (6)

26. State **two** reasons why recycling is important.   (2)

27. The diagram below shows the carbon cycle.

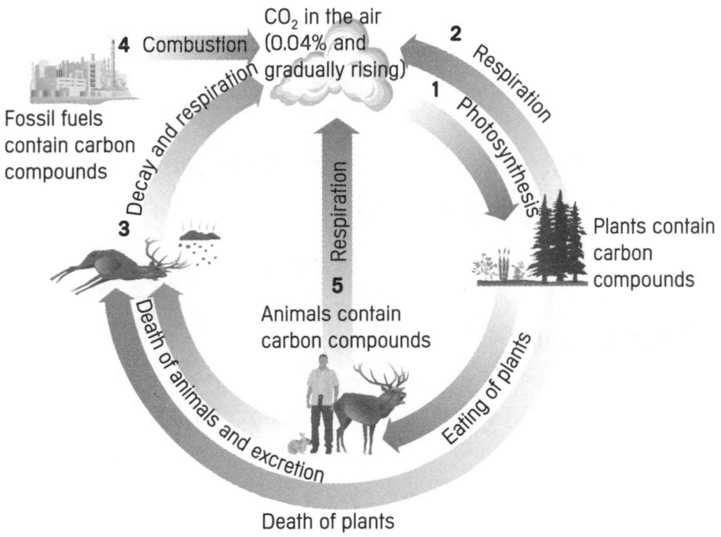

(a) State how carbon dioxide is removed from the atmosphere.   (1)

(b) Describe the process by which the carbon from atmospheric carbon dioxide becomes part of a carbohydrate molecule in the body of a consumer. (2)

(c) Describe the **two** processes by which the carbon in organisms is eventually returned to the atmosphere as carbon dioxide.

(i) (2)

(ii) (2)

(d) Some carbon can be removed from the cycle for long periods of time (from hundreds to millions of years). State **two** possible reasons for this. (2)

(e) Explain why it is important for the balance of the carbon cycle to be maintained. (1)

(f) State **one** human activity that generates large volumes of carbon dioxide and is therefore starting to upset the balance of the carbon cycle. (1)

28. Describe the steps in eutrophication. (4)

(Total: _____ / 129)

Higher Tier

29. Gemma reads a newspaper about 'superbugs' in her local hospital. She is going in for an operation next week and is scared by what she reads.

(a) (i) State what superbugs are. Name **one** in your answer. (2)

(ii) Explain why they have come about. (1)

(b) Explain why they are a particular problem in hospitals. (1)

30. Describe what mutualism is. State **two** examples of it in your answer. (3)

31. Alexander owns a farm on which he rears beef cattle.

(a) Explain how nitrates in the soil help Alexander's grazing cattle to grow. (2)

(b) Describe how nitrogen compounds in cattle can be transferred to other organisms. (3)

continued...

(c) Alexander's farm uses organic practices. The manure (waste products) from the cattle is spread on the fields. Describe how the nitrogen in the manure can be returned to the atmosphere. (3)

32. When Alexander first got his farm, he had one field that he used for silage (grass that is cut and stored for the cows to eat over winter). He cut the grass in this field for silage for several years and did nothing else with the land.

(a) Describe what you would expect to happen to the growth of the grass in that field over time. (1)

(b) State **two** things that Alexander could have done to prevent this.

(i) (1)

(ii) (1)

33. Describe how nitrogen is retained within the nitrogen cycle for long periods of time. (1)

(Total: .... / 19)

Questions labelled with an asterisk (*) are ones where the quality of your written communication will be assessed – you should take particular care with your spelling, punctuation and grammar, as well as the clarity of expression, on these questions.

1. State the difference between the way light and electron microscopes work. In your answer, identify which is more powerful. (3)

    ....

    ....

    ....

*2. Describe the similarities and differences between plant and animal cells. (6)

    ....

    ....

    ....

    ....

    ....

    ....

    ....

    ....

    ....

    ....

3. **(a)** State the functions of cell walls and cell membranes. (2)

    ....

    ....

    (b) State where photosynthesis takes place in plant cells. (1)

    ....

    (c) Describe vacuoles in plant cells. (1)

    ....

    ....

4. Mitochondria are organelles found in most plant and animal cells, and are the major site of respiration.

(a) State **one** example of a type of cell likely to have many mitochondria. (1)

....

(b) Explain your choice. (1)

....

....

5. State how plasmid DNA is different from chromosomal DNA in bacteria. (1)

....

....

6. Label the parts of this bacterial cell. (4)

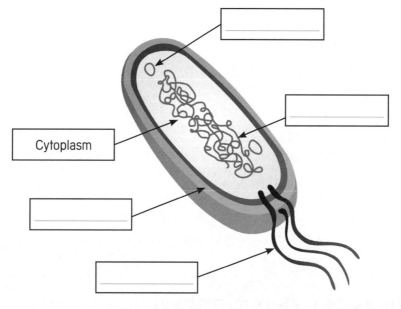

Cytoplasm

7. Chromosomes consist of long, coiled molecules of DNA.

(a) State the **two** types of human cells that do not have 23 pairs of chromosomes. (1)

....

(b) Explain why these cells only have 23 chromosomes. (1)

....

....

8. **(a)** State what genes and chromosomes are made from. (1)

....

(b) State the difference between genes and chromosomes. (2)

....

....

....

....

9. The structure of DNA is most accurately described as a: (1)

A ◯ double helix

B ◯ helix

C ◯ ladder

D ◯ twisted column.

10. A DNA molecule consists of two strands linked together. Describe, with the help of a diagram, the structure of DNA. (4)

....

....

....

....

....

11. The DNA base cytosine pairs with: (1)

A ◯ adenine

B ◯ thymine

C ◯ insulin

D ◯ guanine.

*12. Describe the process for extracting DNA from plant cells. (6)

13. **(a)** Name the **two** scientists that discovered the structure of DNA. (2)

(b) Name the **two** scientists that helped them in this discovery. (2)

(c) **(i)** Name the **one** scientist of the four that was not awarded the Nobel prize. (1)

(ii) Explain why this scientist was not awarded the Nobel prize. (1)

14. **(a)** State what genetic engineering is. (1)

(b) Describe **one** way in which bacteria have been altered in this process. (1)

15. Describe **two** applications of genetic engineering. For each example, state the organism into which the gene was inserted. Explain the benefits of this genetic engineering in your answer. (4)

16. **(a)** 'Sexual reproduction promotes variation.' Explain this sentence. (3)

(b) State how asexual reproduction is different from sexual reproduction in terms of inheriting genes. (2)

17. The following statements refer to mitosis. Draw a diagram to illustrate each stage. (4)

(a) Parent cell with two pairs of chromosomes

(b) Each chromosome replicates itself

(c) The copies separate and the cell divides

(d) Genetically identical daughter cells are formed

18. Draw straight lines to correctly match up mitosis and meiosis with **two** statements for each. (4)

		Involved in sexual reproduction
Mitosis		A diploid cell divides to produce more diploid cells
Meiosis		Produces genetically different cells
		Needed for growth and repair

19. Describe why meiosis must occur before fertilisation. (2)

....

....

....

20. How many daughter cells does meiosis produce? (1)

A ◯ One

B ◯ Two

C ◯ Four

D ◯ Eight

*21. Describe the differences between mitosis and meiosis. (6)

....

....

....

....

....

....

....

....

....

22. **(a)** State what a **clone** is. (1)

(b) Describe **three** of the ethical concerns about cloning. (3)

23. **(a)** State what a **stem cell** is. (1)

(b) Describe how plant and animal stems cells differ in terms of differentiation. (2)

(c) Explain why scientists think that stem cells could be very useful in the field of medicine. (1)

24. **(a)** State **three** types of specialised cells that stem cells have the potential to develop into. (3)

(b) What has to be added to stem cells to make them differentiate? (1)

25. **(a)** Explain why Parkinson's disease develops. (1)

(b) Describe how stem cell research could lead to a cure for Parkinson's disease. (2)

26. Stem cell research is controversial. Suggest **two** opposing views surrounding it. (2)

27. **(a)** What molecules are made following the instructions in DNA? (1)

(b) State **two** reasons why the body needs these molecules. (2)

28. **(a)** Describe what a DNA mutation is. Give **two** ways in which it can occur in your answer. (3)

(b) In terms of inheritance, are mutations always harmful? Explain your answer. (2)

29. **(a)** State what enzymes are also known as. (1)

(b) What do enzymes control? (1)

(c) State the general name for substances that enzymes act upon. (1)

(d) State **two** processes in the human body that involve enzymes. (2)

(e) Describe what denatured means, in terms of enzymes. (1)

30. Describe, with the help of a diagram, how the lock and key hypothesis models how enzymes work. (4)

....

....

....

....

....

....

....

....

....

....

31. Below is a graph that shows the effects of temperature on enzyme action.

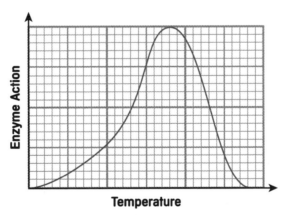

(a) Label the optimum temperature and maximum enzyme activity. (2)

(b) Describe what happens to the activity of the enzyme as the temperature increases up to the optimum temperature. (1)

....

(c) Describe what happens to the activity of the enzyme as the temperature increases beyond the optimum temperature. (1)

....

32. State the temperature at which enzymes in the human body work most effectively. (1)

....

33. **(a)** Draw a graph of how changes in pH affect enzyme action. (4)

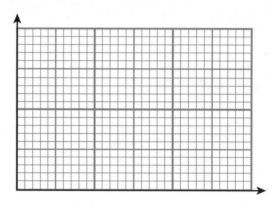

(b) Describe how changing pH affects enzyme action. (4)

....

....

....

(c) The graphs below show the action of two enzymes in the human body. Suggest which enzymes they could represent. (2)

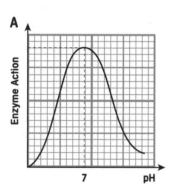 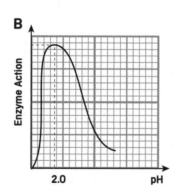

....

....

....

....

....

(Total: .... / 121)

Higher Tier

34. **(a)** State what a **genome** is. (1)

....

(b) (i) Describe what the Human Genome Project achieved. (1)

....

continued...

(ii) Suggest **two** uses that might arise from the Human Genome Project. (2)

....

....

....

....

*35. Describe the stages in cloning Dolly the sheep. (6)

....

....

....

....

....

....

....

*36. Describe the processes of transcription and translation in protein synthesis. (6)

....

....

....

....

....

....

....

....

(Total: .... **/ 16)**

Questions labelled with an asterisk (*) are ones where the quality of your written communication will be assessed – you should take particular care with your spelling, punctuation and grammar, as well as the clarity of expression, on these questions.

1. This is the word equation for the chemical reaction that takes place during aerobic respiration.

    Glucose   +   Oxygen ⟶ Carbon Dioxide   +   Water

    (a) State the balanced symbol equation for this reaction. (2)

    ....

    (b) State what else is produced during aerobic respiration. (1)

    ....

    (c) Describe where the glucose and oxygen come from and how they get to respiring cells. (4)

    ....

    ....

    ....

    ....

    ....

    ....

    (d) State the name of the process by which reactants and products of aerobic respiration enter and leave the cells. (1)

    ....

    (e) How does the body remove the **two** waste products of aerobic respiration? (2)

    ....

    ....

    ....

    (f) Describe what happens to the amount of oxygen and carbon dioxide entering and leaving a muscle cell when you exercise. (2)

    ....

    ....

    ....

Edexcel GCSE Biology Workbook Answers

Model answers have been provided for the quality of written communication questions that are marked with an asterisk (*). The model answers would score the full 6 marks available. If you have made most of the points given in the model answer and communicated your ideas clearly, in a logical sequence with few errors in spelling, punctuation and grammar, you would get 6 marks. You will lose marks if some of the points are missing, if the answer lacks clarity and if there are serious errors in spelling, punctuation and grammar.

B1 Classification, Variation and Inheritance (Pages 4–16)

1. **(a)–(e) In any order:** Plants; Animals; Fungi; Protoctista; Prokaryotes
*2. Plants are multicellular, have cell walls, have chlorophyll and feed autotrophically. Animals are multicellular, do not have cell walls, do not have chlorophyll and feed heterotrophically. Fungi are multicellular or unicellular, have cell walls, do not have chlorophyll and feed saprophytically. Protoctista are unicellular and have a nucleus. Prokaryotes are unicellular and do not have a nucleus.
3. Colour is not specific enough to be a useful method of classification.
4. **Accept any suitable answers, e.g.** Is it warm or cold blooded? Does it have lungs or gills? Does it lay eggs or give birth to live young?
5. More similarities between the organisms as the groups get smaller
6. **(a) (i) Accept any suitable answer, e.g.** Duck–billed platypus
      (ii) Accept any suitable answer, e.g. Because it has fur like a mammal, but a bill and lays eggs
    (b) Scientists are not sure if viruses are alive **(1 mark)** because they do not complete all of the 7 life processes **(1 mark)**
7. Vertebrates have a backbone; Invertebrates do not have a backbone
*8. Fish are cold-blooded, have gills, lay eggs in water and have wet scales. Amphibians are cold-blooded. The adults have lungs (young have gills), lay eggs in water or damp places and have smooth, moist and permeable skin. Reptiles are cold-blooded, have lungs, lay eggs and have dry scaly skin. Birds are warm-blooded, have lungs, lay eggs and have feathers and a beak. Mammals are warm-blooded, have lungs, give birth to live offspring, produce milk and have hair or fur.
9. Annelids (segmented worms) have bodies that are divided into segments and each segment has bristles **(1 mark)**; Molluscs have bodies that are unsegmented, are variable in shape and some have shells **(1 mark)**; Crustaceans have bodies that are divided into three regions, breathe through gills, have an external skeleton and jointed legs **(1 mark)**; Arachnids have bodies that are divided into two regions, four pairs of legs, an external skeleton and jointed legs **(1 mark)**; Insects have bodies that are divided into three regions, three pairs of legs, wings, an external skeleton and jointed legs **(1 mark)**
10. A Scorpion; B Spider; C Lobster; D Bee; E Woodlouse
11. B **should be ticked**.
12. **(a) 1 mark for each characteristic; 1 mark for the reason, e.g.** Cacti have a thick, waxy surface to reduce water loss, spines for protection from predators, stomata that only open at night to reduce water loss and shallow-spreading or deep roots to absorb water; Camels have a large surface area to volume ratio to increase heat loss, body fat stored in a hump, sandy-coloured coat for camouflage, lose little water through urine and sweat and can drink up to 120 gallons of water at any one time
      (b) Accept any two suitable answers, e.g. Hydrothermal vents on the ocean floor; The Antarctic
13. **Accept any two suitable answers for each type of organism, e.g.** Both predators and prey have keen senses to hunt or avoid being hunted; Predators are often able to move very quietly to avoid detection whilst hunting; Prey are often camouflaged to avoid predation; Predators often have eyes on the front of their heads to judge distance; Prey often have eyes on the sides of their heads to allow them to see predators more easily
14. **(a)** The differences between individuals of the same species
      (b) Accept any suitable answers, e.g. Skin and eye colour; The length and shape of noses
15. **In either order:**
      (a) Factor: genes. Example: eye colour (or alternative)
      (b) Factor: environment. Example: language spoken (or alternative)
16. **(a)** Combination
      (b) Genetic
      (c) Environmental
17. It is generally thought that a person's natural ability in sport is genetic **(1 mark)**; Environmental factors such as good coaching, positive support and good facilities can make you better **(1 mark)**; They cannot make you excel at sport if you do not have the natural ability **(1 mark)**
18. **(a)** Discontinuous: eye colour, blood type (or alternative) Continuous: height, weight (or alternative)
      (b) In either order: Discontinuous variation is shown in bar charts. Continuous variation is shown in line graphs.
      (c) (i) (1 mark for axis, 1 mark for labels, 1 mark for points correctly plotted)

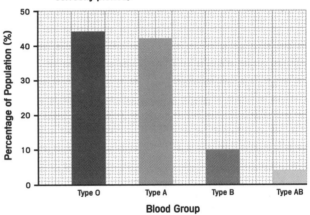

      (ii) The majority of the population have either Type O (44%) or Type A blood (42%) **(1 mark)**; Type B blood is the next most frequent (10%), although far fewer people possess it **(1 mark)**; Very few people possess Type AB blood (4%) **(1 mark)**
      (d)

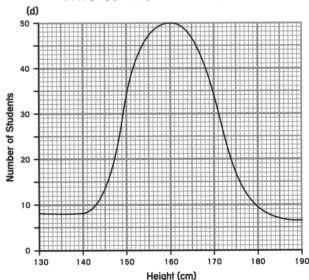

19. (a) **In any order:** Writing in scientific journals **(1 mark)**; Peer-reviewing each other's work **(1 mark)**; Presenting their work at scientific conferences **(1 mark)**

(b) C **should be ticked.**

*20. Most populations of organisms have individuals which vary slightly. Most organisms produce more young than will survive to adulthood. Population sizes are generally stable and so there must be competition between organisms. Those with advantageous characteristics are more likely to survive. Better adapted organisms are more likely to reproduce and pass on their characteristics. Over time, more individuals in a population will possess the advantageous characteristics.

21. (a) The imprints or remains of plants and animals from millions of years ago, preserved in sedimentary rocks

(b) Not all fossils have been discovered yet **(1 mark)**; Some body parts are not fossilised **(1 mark)**; Fossilisation does not always occur **(1 mark)**

22. Numerous small changes over a long period of time have produced slightly different organisms, each adapted for survival in slightly different conditions **(1 mark)**; Some of these have eventually evolved into humans, whilst others evolved into different types of apes **(1 mark)**

23. (1) Cell
(2) Nucleus
(3) Chromosome
(4) Gene

24. (a) Sections of DNA that code for a particular inherited characteristic, e.g. eye colour.
(b) Long-coiled molecules of DNA
(c) 23 pairs
(d) **Accept either:** How our cells function; What characteristics we have

25. (a) Alternative forms of the same gene
(b) Recessive

26. (a) **Accept any suitable answers, e.g.** Not being able to digest food properly **(1 mark)**; Airways becoming clogged with excess, sticky mucus **(1 mark)**
(b) **Accept any suitable answers, e.g.** Great pain in areas where blood cannot reach **(1 mark)**; Shortness of breath **(1 mark)**

27. (a) Homozygous (dominant)
(b) Homozygous (recessive)
(c) Heterozygous

28. (a) Genotypes are letters that scientists use to simplify alleles
(b) Phenotypes are the words that represent the physical expression of genotypes

29. (a) Heterozygous
(b) Homozygous dominant
(c) Homozygous recessive

30. **In any order:** EE unattached lobes **(1 mark)**; Ee unattached lobes **(1 mark)**; ee attached lobes **(1 mark)**

31.

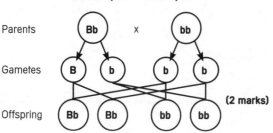

(2 marks)

(2 marks)

32. If both parents are heterozygous (i.e. have one dominant allele and one recessive allele) there is a 25% chance that the offspring will inherit two recessive alleles (and therefore, blue eyes) **(2 marks)**

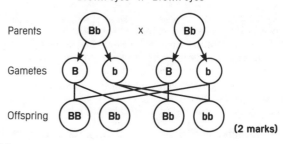

(2 marks)

33. (a)

	E	e			e	e			e	e
E	EE	Ee		**E**	Ee	Ee		**e**	ee	ee
e	Ee	ee		**e**	ee	ee		**e**	ee	ee

(b) (i) 75%   **(ii)** 50%   **(iii)** 0%

34. 100% offspring will have brown eyes; 50% of the offspring will be homozygous dominant and the other 50% heterozygous **(2 marks)**

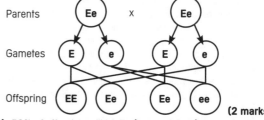

(2 marks)

35. (a) 75% of offspring will have unattached lobes and 25% won't; 25% will be homozygous dominant, 50% heterozygous and 25% homozygous recessive **(2 marks)**

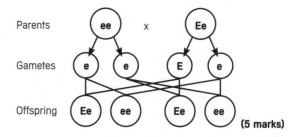

(2 marks)

(b) 50% of offspring will be Ee (heterozygous) and so will have unattached lobes, 50% of offspring will be ee (homozygous recessive) and so will have attached lobes

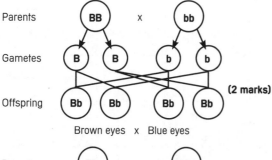

(5 marks)

36. 100% of offspring will have unattached lobes; 50% of the offspring will be homozygous dominant and the other 50% heterozygous

Unattached lobes x Unattached lobes

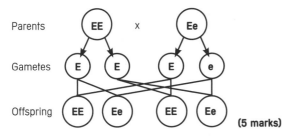

Parents

Gametes

Offspring

(5 marks)

37. (a) They have two different species as parents **(1 mark)**; They are fertile **(1 mark)**; This makes the classification of them difficult **(1 mark)**
   (b) (i) Overlapping populations of different species of closely related organisms
   (ii) They are able to interbreed only in the overlapping populations
   (iii) Accept any suitable answer, e.g. Salamanders on the west coast of the USA
38. Two names (in Latin)
39. It allows habitats of species that need to be conserved to be identified
40. Sickle cell disease is inherited when an individual receives a 'faulty' recessive gene from each parent **(1 mark)**; If a person inherits only one recessive gene they are a 'carrier' which means they do not have the disorder but could pass it to their children **(1 mark)**
41. (a) The process of evolution by which new species form.
   (b) Geographical features separate populations which then evolve separately
   (c) Accept any suitable answer, e.g. The three-spined stickleback

B1 Responses to a Changing Environment (Pages 17–25)

1. (a) The maintenance of a stable internal environment
   (b) Accept any two from: Keeping the correct levels of water in our bodies; Keeping a constant temperature; Keeping blood glucose levels constant
   (c) (i) Because organisms work most effectively in the conditions at which homeostasis returns them to
   (ii) If organisms are unable to return to these conditions they may die
2. It becomes more concentrated (and less in volume) **(1 mark)**; It appears darker yellow **(1 mark)**
3. Blood vessels in the skin restrict, reducing heat loss **(1 mark)**; Muscles start to shiver, causing heat energy to be released **(1 mark)**; Erector muscles cause hairs to stand up, trapping heat **(1 mark)**; Sebaceous glands produce oily sebum to waterproof the skin from water **(1 mark)**
4. (a) 37°C
   (b) Because the enzymes in the human body work best at this temperature
   (c) Hypothalamus
5. (a) Because it allows organisms to react to their surroundings and coordinate their behaviour.
   (b)

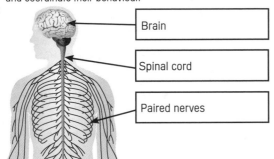

Brain

Spinal cord

Paired nerves

6. (b) Sight
   (c) Touch
   (d) Taste
   (e) Smell
7. Cerebellum: coordinates movement and balance
   Medulla: controls automatic actions like heart beat and breathing
   Frontal lobe: controls higher mental functions like choice and memory
   Cerebral hemisphere: responsible for numerical computation, language and emotions
8. (Award 1 mark per cell only if arrows are in the correct direction)

Sensory neurone

Relay neurone

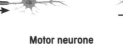

Motor neurone

9. It is elongated to make fast connections from one part of the body to another **(1 mark)**; Branched endings connect a single neurone with many muscle fibres **(1 mark)**; The cell body connects to many other neurones **(1 mark)**
10. (a)

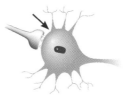

   (b) A synapse
   (c) An impulse travels down neurone A to the synapse **(1 mark)**; A chemical neurotransmitter is released across the gap **(1 mark)**; The neurotransmitter activates receptors on neurone B **(1 mark)**; This causes an impulse to be transmitted down neurone B **(1 mark)**
11. (b) 3
   (d) 6
   (e) 4
   (f) 5
12. (a) A voluntary response is one that we have complete control of **(1 mark)**; (e.g. speaking, walking and picking something up or alternative) **(1 mark)**; Involuntary responses happen automatically **(1 mark)**; (e.g. blinking and moving parts of your body away from pain or alternative) **(1 mark)**
   (b) Accept any two suitable answers, e.g. To investigate involuntary responses you could shine a light in someone's eye to see the pupil reflex **(1 mark)**; You could also cool the skin on their arm and observe what happens to the hairs there **(1 mark)**
13. (a) Because they protect the body from harm
   (b) A: Sensory neurone
   B: Relay neurone
   C: Motor neurone
   D: Muscle (effector)
   (c) A receptor in the skin detects heat (stimulus), which sends impulses down a sensory neurone **(1 mark)**; In the spinal cord, the sensory neurone synapses with a relay neurone, bypassing the brain **(1 mark)**; The relay neurone synapses with a motor neurone, sending impulses to muscles (effectors) in the arm **(1 mark)**; The muscles contract and pull away from the hot object **(1 mark)**
*14. Use an unwound paperclip (or similar two pointed object) to touch two points on your partner's skin. Start reasonably far apart and move the points closer together. Stop when they can only feel one point and measure this distance. Repeat this process on different parts of the skin. A closer distance between the two points indicates more touch receptors in that area. Lips and fingertips have receptors very close together.

15. **(a)** Hormones
    (b) Endocrine glands
    (c) In the blood
    (d) (i)–(iii) **Accept any three from:** Adrenaline produced in the adrenal glands; FSH and LH produced in the pituitary gland; Testosterone produced in the testes; Oestrogen and progesterone produced in the ovaries; Insulin produced in the pancreas

16. **(a)** D **should be ticked**.
    (b) A medical condition where the amount of glucose in the blood cannot be lowered.

17. The pancreas releases insulin **(1 mark)**; Glucose from the blood is converted into glycogen in the liver **(1 mark)**; This removes it from the blood and lowers blood glucose levels **(1 mark)**

18. **(a) Accept any three from:** Urinating more often than normal; Becoming more thirsty; Increased tiredness; Weight loss; Blurred vision
    (b) (i) unable to produce insulin **(1 mark)**
        (ii) resistant to the insulin they produce **(1 mark)**
    (c) Accept any three suitable answers, e.g. Eat three meals a day **(1 mark)**; Include some carbohydrate but reduce fat and sugar in the diet **(1 mark)**; Be physically active **(1 mark)**

19. **(a)** Because the blood sugar concentration fluctuates greatly
    (b) Because the person has just eaten breakfast (A) and lunch (B)
    (c) Blood sugar concentration would only rise slightly/less steeply
    (d) Because their blood sugar concentration dropped very low

20. **(a)** Positively phototropic: plant shoots grow towards the light **(1 mark)**; Positively geotropic: plant roots grow downwards in the direction of gravity **(1 mark)**
    *(b) Put cress seeds on damp cotton wool in three small containers. Place one on a window sill to grow normally (the control). Place the second on the same window sill but cover it with a second container to block out the light. Place the third on the same window sill but cover it with a third container to block out the light in all but a small area. Allow the seedlings to grow. The seedlings in the first dish should be tall, dark green and upright. The seedlings in the second dish should be taller, thinner, yellow and pointing in all directions seeking the light. The seedlings in the third dish should be taller, thinner, a little yellower and pointing towards the light.
    (c) (1 mark for light source, 2 marks for suitably sized cells)

Light source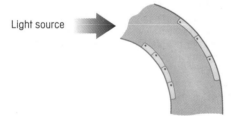

21. Auxins: promote cell elongation **(1 mark)**; Gibberellins: also promote cell elongation but also encourage flowering, break seed dormancy and can be used to force seedless fruit to develop **(1 mark)**

22. **(a)** Diagram B
    (b) Blood flow through the superficial capillaries is reduced (vasoconstriction) **(1 mark)**; This reduces heat loss from radiation and convection **(1 mark)**; Sweating is also reduced **(1 mark)**; This reduces heat loss from evaporation **(1 mark)**
    (c) Blood flow through the superficial capillaries is increased (vasodilation) **(1 mark)**; This increases heat loss from radiation and convection **(1 mark)**; Sweating is also increased **(1 mark)**; This increases heat loss from evaporation **(1 mark)**

23. The pancreas releases glucagon **(1 mark)**; Glycogen from the liver is converted into glucose **(1 mark)**; This is released into the blood and raises blood glucose levels **(1 mark)**

24. **Accept any three from:** Selective weedkillers to disrupt the growth of plants other than grass; Rooting powder to encourage cuttings to form roots; Fruit ripening to allow immature fruit to be picked and transported to shops

B1 Problems of, and Solutions to a Changing Environment (Pages 26–36)

1. **(a)** It speeds up the transmission of a nervous impulse across a synapse
   (b) Accept any two suitable answers, e.g. The body reacts quicker, but it can keep you awake, leading to physical exhaustion **(1 mark)**; It can also make you tense/edgy **(1 mark)**

2. It prevents the transmission of an impulse across a synapse

3. They are highly addictive

4. **Accept any suitable answer, e.g.** The risk of contracting HIV or hepatitis from an unsterilised needle

5. **In any order:**
   (a) Both are natural drugs, obtained from plants
   (b) Both can be used to relieve pain
   (c) Both are illegal
   Accept any other suitable answer.

6. **(a)** Nicotine
   (b) Tar
   (c) (i)–(iii) Accept any suitable answers, e.g. Emphysema; Bronchitis; Heart disease

7. **(a) Accept any two suitable answers, e.g.** Alveoli in the lungs are damaged **(1 mark)**; This reduces the surface area for gaseous exchange **(1 mark)**; Cilia in the airways stop beating **(1 mark)**; This results in a build up of mucus **(1 mark)**
   (b) Blood vessels are damaged **(1 mark)**; This can lead to strokes and heart attacks **(1 mark)**; Blood pressure is increased **(1 mark)**; This can lead to strokes and heart attacks **(1 mark)**

8. **(a) Accept any two suitable answers, e.g.** Slows reactions; Depresses brain function; Loss of self-control
   (b) In either order: Liver damage **(1 mark)**; Brain damage **(1 mark)**

9. **(a)** Solvents are liquids **(1 mark)**; that are used to dissolve glue, paint, etc. **(1 mark)**
   (b) Accept any two from: Slows reactions; Hallucinations; Behavioural changes; Damage to liver, lungs, brain, or kidneys

*10. Sit at a table with your arm resting on it and your open hand off the edge. Allow your partner to line up the bottom of a ruler with your thumb and first finger. Catch the ruler when they drop it without telling you. Measure the length of the ruler from the bottom to the point which you are holding. Convert this distance into a reaction time. Repeat and take an average reading to make it reliable. Drink a caffeinated drink and wait for a short period of time for the caffeine to be absorbed into your blood. Repeat the experiment and compare the results.

11. When a tissue or organ is moved from one patient to another or from one part of a patient to another **(1 mark)**; **Accept any two examples from**: Heart; Kidney; Liver; Lung **(2 marks)**

12. Some people believe that we should be forced to give up our organs upon death (not volunteer) **(1 mark)**. Others believe that people who may have had some control over their condition should not be allowed transplants (e.g. liver transplants for alcoholics) **(1 mark)**. Organs are sometimes stolen and trafficked **(1 mark)**

13. **(a)** A disease-causing microorganism
    (b) Accept any two suitable answers, e.g. Bacteria: e.g. Cholera or Salmonella; Fungi: e.g. athlete's foot or ringworm; Virus: e.g. Influenza or HIV

14. **(a)** Indirect contact is most common.
    (b) Accept any two suitable answers, e.g. Dirty drinking water: Cholera; Uncooked food: Salmonella; Airborne: Influenza; Direct contact: Athlete's foot; Body fluids: HIV; By a vector: Malaria
    (c) Another organism that acts as a vehicle or carrier of a pathogen **(1 mark)**; **Accept any suitable example, e.g.** Mosquitoes are the vector that spreads malaria **(1 mark)**

15. Skin: Prevents infection from pathogens; Physical barrier **(1 mark)**
    Cilia: Beat rhythmically to remove mucus containing pathogens; Physical barrier **(1 mark)**
    Lysozymes: Enzymes in tears which destroy pathogens; Chemical barrier **(1 mark)**

16. **Accept any suitable answer, e.g.** Taxol **(1 mark)** from the Pacific Yew Tree **(1 mark)**

17. (a) Antifungals kill fungi **(1 mark)**; Antibacterials kill bacteria **(1 mark)**
    (b) Viral infections
18. (a) Substances that are spread on surfaces (often skin) to prevent infection by pathogens
    *(b) Divide the area of an agar plate covered with bacteria into sections. Soak disks of filter paper with different antiseptics. Soak a disk in water (the control). Put one disk in each section of the agar plate and incubate for several days. Measure the distance of clearing around each disk. The larger the clearing is, the more effective the antiseptic.
19. The dynamic relationship between all living things
20. (a) **Accept either answer:** The feeding relationship between organisms; The transfer of energy and biomass
    (b) The transfer of energy
    (c) A diagram which shows the number of each organism present in each stage of a food chain
    (d)

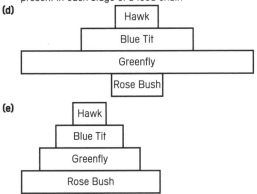

    (f) A diagram which shows how energy and biomass are lost at each stage, as they are transferred up a food chain
    (g) Pyramids of number sometimes have a large bar at the top if parasites are included **(1 mark)**; They also sometimes have a small bar at the bottom if the producer is a single plant (here, a rose bush) **(1 mark)**
21. (a) Photosynthesis
    (b) **Accept any suitable answer, e.g.** Respiration
    (c) 0.1%
    (d) **Answer should make reference to:** Movement; Excretion; Mating; Keeping warm
22. (a) Parasites
    (b) **Accept any two suitable answers, e.g.** Fleas; Lice
23. (a)

Population (Billions) vs Date graph showing population rising sharply near 2000 CE, with x-axis from 8000 BCE through 2000 CE

    (b) **Accept any two suitable answers e.g.** Extinctions; Global warming; Pollution
24. (a) **Accept any two from:** Hydrocarbons released from the burning of fossil fuels; Carbon dioxide released from the burning of fossil fuels; Sulfur dioxide released from the burning of fossil fuels; Carbon monoxide released from vehicle exhausts **(2 marks for each correct answer)**
    (b) **Accept any two from:** Sewage from human waste; Nitrates from fertilisers; Phosphates from fertilisers **(2 marks for each correct answer)**
    (c) (i) Indicator species
        (ii) **Accept any two suitable answers, e.g.** Stonefly; Freshwater shrimps

*25. Place some seeds on damp cotton wool in three Petri dishes. Apply a solution of ammonium nitrate and ammonium phosphate to the cotton wool in two dishes. Leave the third dish untreated (the control). Leave all three dishes in the same sunny area for a few days. Compare the growth of the seedlings in the control dish with those treated with the pollutants.
26. **Accept any two suitable answers, e.g.** Reduces waste; Saves energy; Protects the environment from mining, logging, etc.
27. (a) By photosynthesis
    (b) Green plants take in carbon dioxide to produce glucose **(1 mark)**; Animals feed on these plants **(1 mark)**
    (c) **In either order:**
        (i) Respiration: plants and animals respire releasing carbon dioxide **(2 marks)**
        (ii) Decay: when plants and animals die, other animals and microorganisms feed on their bodies, causing them to break down and release carbon dioxide into the air **(2 marks)**
    (d) Carbon dioxide absorbed from the atmosphere by trees becomes 'locked up' as wood **(1 mark)**; When the tree dies, it takes millions years to turn into coal **(1 mark)**
    (e) Because increasing the amount of carbon dioxide released into the atmosphere may be the cause of global warming
    (f) **Accept any suitable answer, e.g.** Burning fossil fuels; Deforestation
28. Excess fertiliser is washed into streams and rivers and accumulates in ponds and lakes **(1 mark)**; Nitrates cause excessive algal growth which blocks out sunlight to other plants **(1 mark)**; The other plants cannot photosynthesise so they die and start to rot **(1 mark)**; The rotting process uses up all of the oxygen and so the water cannot support life **(1 mark)**
29. (a) (i) 'Superbugs' are bacteria which have evolved to be immune to some antibiotic medicines **(1 mark)**; **Accept any suitable answer, e.g.** MRSA
        (ii) They have evolved immunity due to misuse or overuse of antibiotics
    (b) Because some patients have weakened immune systems.
30. A relationship between living things where both organisms benefit **(1 mark)**; **Accept any two suitable answers, e.g.** Oxpecker birds that pick ticks from the hides of large animals; Nitrogen-fixing bacteria that live in the roots of legumes **(2 marks)**
31. (a) Plants absorb the nitrates from the soil **(1 mark)**; When the cattle eat the plants, the nitrogen becomes animal protein **(1 mark)**
    (b) When the cattle excrete, nitrogen compounds are passed into the soil **(1 mark)**; Nitrifying bacteria then convert nitrogen compounds into nitrates in the soil **(1 mark)**; The nitrates are taken up by plants which are eaten by other organisms **(1 mark)**
    (c) Soil bacteria convert the manure into ammonia **(1 mark)**; Nitrifying bacteria then convert the ammonia into nitrates in the soil **(1 mark)**; Denitrifying bacteria convert the nitrates into atmospheric nitrogen **(1 mark)**
32. (a) The grass would stop growing so quickly or so well.
    (b) (i)–(ii) **Accept any two suitable answers, e.g.** He could stop using it for silage and allow cattle to graze it for a year; He could use fertiliser to replace nitrates in the soil
33. Nitrogen absorbed from the soil by trees becomes 'locked up' as wood

B2 The Building Blocks of Cells (Pages 37–47)

1. Light microscopes rely on refraction of light to magnify images **(1 mark)**; Electron microscopes use beams of electrons, which have a shorter wavelength than light **(1 mark)**; Electron microscopes are therefore more powerful **(1 mark)**
*2. Plant and animal cells both possess a nucleus. Plant and animal cells both possess a cell membrane. Plant and animal cells both have a cytoplasm. Only plant cells possess a cell wall. Only plant cells possess a vacuole. Only plant cells possess chloroplasts.

3. (a) Cell walls provide structural support **(1 mark)**; Cell membranes control the movement of substances in and out of cells **(1 mark)**
   (b) In the chloroplasts in the green parts of plants
   (c) Large spaces in the centre of cells which are full of sap
4. (a) **Accept any suitable answer, e.g.** Sperm; Muscle; Liver
   (b) Because they are particularly active and need more energy generated from respiration
5. Plasmid DNA is a separate, small circular section of DNA that can replicate independently of the chromosomal DNA.
6.

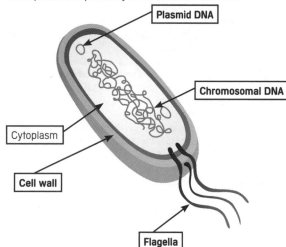

Plasmid DNA
Chromosomal DNA
Cytoplasm
Cell wall
Flagella

7. (a) Sperm and eggs
   (b) When they meet at fertilisation they join to form a cell with 23 pairs
8. (a) DNA
   (b) A small section of DNA that codes for a particular protein is called a gene **(1 mark)**; Many genes join together to form a chromosome **(1 mark)**
9. A **should be ticked**.
10. Two strands are held together by paired bases **(1 mark)**; The bases are adenine, thymine, cytosine and guanine **(1 mark)**; Adenine links to thymine, and cytosine links to guanine **(1 mark)**

(1 mark)

11. D **should be ticked**.
*12. Dissolve 3g of salt in 90cm³ of distilled water in a beaker and mix in 10cm³ of washing up liquid. Mash 50g of defrosted frozen peas in a second beaker with a little water and add 10cm³ of the solution to the first beaker. Place in a water bath at 60°C for 15 minutes and stir gently. Place in an ice bath for 5 minutes and stir gently. Very carefully pour chilled ethanol solution onto the top of the beaker and leave for a few minutes. The DNA will separate out into pale, bubble-covered strands at the boundary between the pea extract and the ethanol.
13. (a) James Watson and Francis Crick
   (b) Rosalind Franklin and Maurice Wilkins
   (c) (i) Rosalind Franklin
       (ii) She died before the other three scientists were awarded it (At this time the prize could only be awarded to the living)
14. (a) The removal of a gene from one organism and insertion into another
   (b) **Accept any suitable answer, e.g.** Bacteria have been altered to include the human gene for insulin
15. **Accept any two suitable answers, e.g.** The beta-carotene gene has been inserted into rice to reduce Vitamin A deficiency in humans; The human gene for insulin has been inserted into bacteria to produce insulin for diabetics

16. (a) Gametes are produced when a cell divides, which 'shuffles' the genes **(1 mark)**; Gametes fuse randomly, with one of each pair of genes coming from each parent **(1 mark)**; A pair of genes may have the same alleles or different alleles, producing different characteristics **(1 mark)**
   (b) In asexual reproduction, the offspring's genes all come from the one parent, so it is genetically identical to the parent (a clone) **(1 mark)**; In sexual reproduction the offspring's genes come from both parents **(1 mark)**
17. (a) (b) (c) (d)
18.

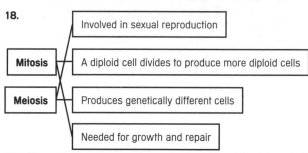

Involved in sexual reproduction
Mitosis — A diploid cell divides to produce more diploid cells
Meiosis — Produces genetically different cells
Needed for growth and repair

19. After meiosis has occurred, the four gamete nuclei only have half the number of chromosomes of the parental cell **(1 mark)**; During fertilisation, two gametes (one from each parent) join so the cell then contains the correct number of chromosomes **(1 mark)**
20. C **should be ticked**.
*21. Mitosis produces two daughter cells, whilst meiosis produces four. Mitosis produces genetically identical daughter cells, whilst meiosis produces different cells. Mitosis produces daughter cells with the same number of chromosomes as the parent cell. Meiosis produces daughter cells with half the number of chromosomes as the parent cell. Mitosis produces new cells for growth, to repair damaged tissues and in asexual reproduction. Meiosis produces gametes (sperm and eggs).
22. (a) An individual that is genetically identical to its parent
   (b) **Accept any three suitable answers, e.g.** The fear of the 'perfect race'; The possibilities of abnormalities occurring in clones; Clones will not have 'parents'; Cloning does not allow 'natural' evolution
23. (a) A cell which has not yet differentiated
   (b) Plant cells can differentiate at any time **(1 mark)**; Animal cells can only differentiate soon after they are made **(1 mark)**
   (c) They could replace damaged cells and tissues and help in the treatment of diseases
24. (a) **Accept any three suitable answers, e.g.** Pancreatic islet cell; Heart muscle cell; Blood cells; Neurones; Bone marrow cell
   (b) Growth factors
25. (a) Brain neurones are faulty (they stop producing dopamine)
   (b) Stem cells develop into neurones **(1 mark)**; If they were inserted into the brain they might produce dopamine and cure Parkinson's disease **(1 mark)**
26. **Accept any two suitable answers, e.g.** Some people think that an embryo is an individual life and so disagree with their use in stem cell research **(1 mark)**; Other people agree with the research and think it might be a cure for conditions like Parkinson's disease **(1 mark)**
27. (a) Proteins
   (b) **Accept any two suitable answers, e.g.** To make enzymes, hormones, skin, hair, etc. and for growth and repair
28. (a) A change to the sequence of bases that make up a gene **(1 mark)**; **Accept any two from:** UV radiation; Viruses; Some chemicals; Errors during DNA replication
   (b) Mutations are not always harmful **(1 mark)**; Some have no effect at all, whilst some are advantageous **(1 mark)**
29. (a) Biological catalysts
   (b) The rate of biological reactions inside living organisms
   (c) Specific substrates
   (d) **Accept any two suitable answers, e.g.** Digestion; Protein synthesis
   (e) It means that the enzyme is irreversibly damaged and will no longer work

30. The enzyme is the lock and the substrate is the key **(1 mark)**; The substrate enters the enzyme (like a key in a lock) **(1 mark)**; It is then broken down **(1 mark)**; The enzyme is specific for the substrate, like a lock and key **(1 mark)**

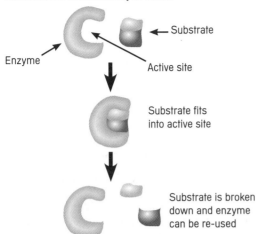

Enzyme · Substrate · Active site

Substrate fits into active site

Substrate is broken down and enzyme can be re-used

31. (a) (1 mark for each label)

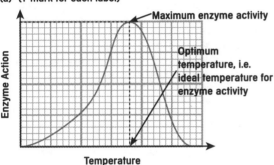

Maximum enzyme activity

Optimum temperature, i.e. ideal temperature for enzyme activity

Enzyme Action — Temperature

(b) It increases
(c) It decreases

32. 37°C

33.

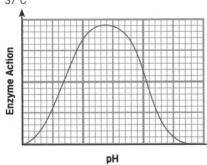

Enzyme Action — pH

(a) 1 mark each for: correct axes; appropriate labelling of axes; appropriately shaped graph; smooth curve
(b) When the pH is low the enzyme action is low **(1 mark)**; As the pH increases so does the enzyme action, until the optimum is reached **(1 mark)**; Further increases in pH result in a decrease in enzyme action **(1 mark)**; There is no enzyme action at extreme high or low pH **(1 mark)**
(c) A: Amylase; B: Protease

34. (a) All the genetic material in an organism
(b) (i) It identified the sequence of all bases in every gene of the human genome
(ii) Accept any two suitable answers, e.g. To identify and replace 'faulty' genes; To compare DNA samples from potential suspects at crime scenes

*35. The diploid nucleus is taken from a mature cell (ordinary body cell) of the donor organism. The diploid nucleus, containing all of the donor's genetic information, is inserted into an empty egg cell (i.e. an egg cell with the nucleus removed or enucleated). This is nuclear transfer. The egg cell, containing the diploid nucleus, is stimulated so that it begins to divide by mitosis. The resulting embryo is implanted in the uterus of a 'surrogate mother'. The embryo develops into a foetus and is born as normal.

*36. Transcription begins when the DNA of the gene that codes for the protein to be copied unzips. mRNA copies the base pair sequence of this gene. mRNA exits the nuclear membrane into the cytoplasm.
Translation then begins when the mRNA attaches to a ribosome. tRNA molecules align opposite the mRNA, each bringing with them an amino acid. The amino acids form to make a polypeptide, and then a protein.

B2 Organisms and Energy (Pages 48–57)

1. (a) 1 mark for reactants, 1 mark for products
$$C_6H_{12}O_6 + 6O_2 \longrightarrow 6CO_2 + 6H_2O$$
(b) Energy
(c) Glucose is obtained from food **(1 mark)**; Oxygen is taken in from the air **(1 mark)**; They are transported in the blood **(1 mark)**; Glucose and oxygen move from the capillaries into the respiring cells by diffusion **(1 mark)**
(d) Diffusion
(e) Carbon dioxide is breathed out **(1 mark)**; Water is lost as sweat, moist breath or excreted as urine **(1 mark)**
(f) More oxygen is absorbed into the muscle cell by diffusion from the capillaries **(1 mark)**; More carbon dioxide is removed by diffusion from the muscle cell into the capillaries **(1 mark)**

2. Respiration is the breakdown of glucose to create energy **(1 mark)**; Ventilation is breathing (This is the process of getting oxygen into and carbon dioxide out of the lungs) **(1 mark)**

3. (a) It increases
(b) It will increase
(c) The breathing rate increases so that more oxygen can be taken in **(1 mark)**; The cells need more oxygen because they are respiring faster **(1 mark)**; This means more carbon dioxide needs to be removed **(1 mark)**; As both these gases are transported in the blood, the heart rate also increases **(1 mark)**

4. (a) 1 mark for reactant, 1 mark for product
Glucose ⟶ Lactic acid
(b) Not enough oxygen can reach her muscle cells, which means she respires anaerobically **(1 mark)**; Glucose is broken down into lactic acid, a waste product which builds up in her muscles and causes them to feel rubbery **(1 mark)**

5. Aerobic respiration

6. (a) When muscle cells have been respiring anaerobically (usually after strenuous exercise)
(b) Lactic acid needs to be broken down into carbon dioxide and water **(1 mark)**; Deep breathing continues after exercise to provide the oxygen to do this **(1 mark)**

7. (a) Accept any two suitable answers, e.g. Leaves have lots of internal air spaces to create a large surface area for gas exchange; Leaves are full of chloroplasts containing chlorophyll for photosynthesis; Leaves are flat to absorb as much sunlight as possible.
(b)

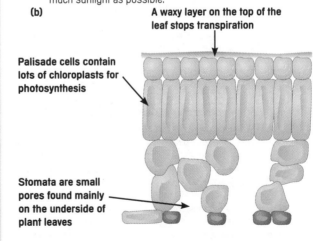

A waxy layer on the top of the leaf stops transpiration

Palisade cells contain lots of chloroplasts for photosynthesis

Stomata are small pores found mainly on the underside of plant leaves

(c) Carbon dioxide + Water $\xrightarrow[\text{chlorophyll}]{\text{light}}$ Glucose + Oxygen
1 mark for reactants, 1 mark for products
(d) Accept any three from: Used in respiration; Stored as starch; Made into cellulose cell walls; Made into proteins

8. **(a)** (Award 1 mark for any line that falls steeply and meets the x-axis between 40 and 45°C

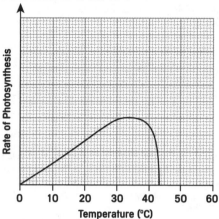

(b) (i) X

(ii)

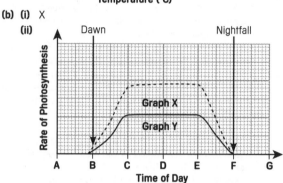

(iii) As the levels of light increased in the morning (between points B and C) so did the rate of photosynthesis **(1 mark)**; At approximately point C the rate reached a maximum and continued until the sun began to set **(1 mark)**; At this point the rate of photosynthesis decreased (between E and F) to zero **(1 mark)**; On a warm, sunny day a higher maximum of photosynthesis rate was observed **(1 mark)**

(c) (i) (1 mark each for: correct axes; appropriately shaped graph; smooth curve)

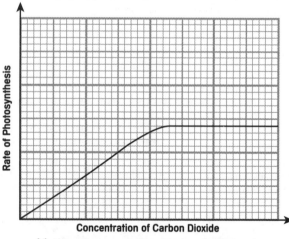

(ii) When the concentration of carbon dioxide was low, so was the rate of photosynthesis **(1 mark)**; As the concentration increased, so did the rate of photosynthesis until a maximum was reached **(1 mark)**; After this, further increases in carbon dioxide had no effect on the rate of photosynthesis **(1 mark)**

9. **(a)** Diffusion is the movement of any substance; Osmosis is the movement of water **(1 mark)**; Osmosis occurs across a partially permeable membrane; diffusion doesn't **(1 mark)**

(b) The sugar solution in the thistle funnel has a lower concentration of water than the beaker **(1 mark)**; As a consequence, water moves from the beaker into the thistle funnel across the Visking tubing **(1 mark)**; This forces the water higher up the tube **(1 mark)**

(c) A and C should be ticked.

10. When the plants are surrounded by salt water there is a higher concentration of water in their cells than the salt water **(1 mark)**; Therefore, water diffuses from their cells causing them to wilt and perhaps die **(1 mark)**

11. **(a)** **Accept any suitable answer, e.g.** Cut out, dry off and measure accurately the mass of some small pieces of potato **(1 mark)**; Place one in a beaker of distilled water for 15 minutes **(1 mark)**; Remove, dry off the potato and measure its mass again **(1 mark)**; Repeat the process using various concentrations of sugar solution **(1 mark)**

(b) The potato in the beaker of distilled water should have gained mass as water entered it by osmosis **(1 mark)**; The potato in the sugar solutions should have lost mass as water left it by osmosis. (Unless one solution is the same concentration as the potato in which case no mass would be lost) **(1 mark)**

12. **(a)** A measure of the variety of different organisms in a habitat

(b) Because sampling entire populations is inefficient or sometimes impossible

(c) Systematic and random

13. **(a)** **Accept any suitable answer, e.g.** They would use quadrats to sample their lawns **(1 mark)**; They could place their quadrat randomly on their lawns and count the numbers of different species of plant that grew in them **(1 mark)**; This would determine the species richness or frequency **(1 mark)**

(b) Make it more reliable

(c) There are many variables that affect how lawns grow besides shade **(1 mark)**; **Accept any two suitable answers, e.g.** The amount of water they receive; The quality of the soil

14. **(a)** Evaporation through pores in their leaves called stomata

(b) Make their stomata smaller

(c) (i) Photosynthesis will be slowed down

(ii) Water is a reactant so less of it means less reaction occurs

*15. Water evaporates from the internal leaf cells through the stomata. Water passes from the xylem vessels to leaf cells due to osmosis. This 'pulls' the entire 'thread' of water in that vessel upwards by a very small amount. Water enters the xylem from the root tissue to replace water which has moved upwards. Water enters root hair cells by osmosis to replace water which has entered the xylem.

16. **(a)** To stop water evaporating from it

(b) It has evaporated from the plant's leaves

(c) (i) There would be less transpiration so less water lost

(ii) There would be less transpiration so less water lost

(iii) There would be more transpiration so more water lost

(iv) There would be more transpiration so more water lost

17. **(a) (i)** Pitfall traps are small containers buried in the ground to catch small animals **(1 mark)**; They sometimes contain food to attract the small animals that fall in and are caught **(1 mark)**

(ii) Sweep netting uses a large net **(1 mark)**; which is swept through undergrowth to catch insects **(1 mark)**

(iii) Kick sampling involves placing a net downstream of an area of stream bed that is gently disturbed **(1 mark)**; to catch animals swept downstream **(1 mark)**

(b) (i) (1 mark for diagram, 1 mark for labels)

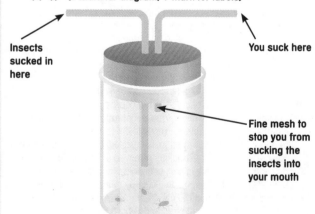

(ii) When one tube is sucked insects are pulled through the other tube into the jar (not the person's mouth)

18. Organisms are caught and marked in a way that will not harm them **(1 mark)**; They are then freed to be caught again during random sampling **(1 mark)**; The number caught that are marked, compared with the number unmarked, can be used to estimate the total population **(1 mark)**

19. **(a) Accept any suitable answer, e.g.** Temperature, pH, light intensity
    (b) Accept any suitable answers, e.g. it can take measurements more frequently or accurately

B2 Common Systems (Pages 58–67)

1. **(a)** The imprints or remains of plants and animals from millions of years ago, preserved in sedimentary rocks
   (b) Accept any two from: Not all fossils have been discovered yet; Some body parts are not fossilised; Fossilisation does not always occur

2. **(a) Specialisation:** the process through which an unspecialised cell becomes a specific type of cell
   (b) Elongation: the process by which cells elongate (stretch out)
   (c) Division: the process by which two cells are formed from one cell (mitosis)

3. **(a)** They do not take into account growth in other directions
   (b) Because dry mass can only be measured when the organism is dead

4. **(a)**

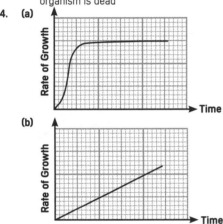

   (b)

   (c) In the first graph, the growth rate is fast to begin with, then it slows down and stops **(1 mark)**; In the second graph, the growth rate is constant **(1 mark)**

5. **In any order:**
   (a) Red blood cells **(b)** White blood cells **(c)** Plasma
   (d) Platelets

6. **(a)** Red blood cells
   (b) (1 mark for each feature, 1 mark for each adaptation)
   Biconcave shape provides a large surface area through which to absorb oxygen; No nucleus so cells can (hold more haemoglobin and) carry more oxygen
   (c) Accept any suitable answer, e.g. In the lungs haemoglobin in the red blood cells combines with oxygen to form oxyhaemoglobin **(1 mark)**; The blood travels round the body through the arteries, delivering the oxygen to tissues/organs **(1 mark)**

7. **(1 mark for diagram, 2 marks for suitable description)**

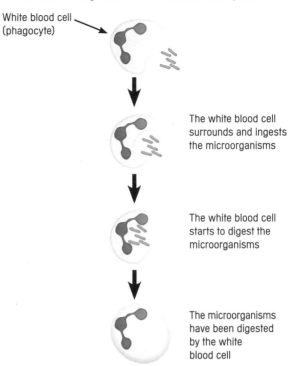

White blood cell (phagocyte)

The white blood cell surrounds and ingests the microorganisms

The white blood cell starts to digest the microorganisms

The microorganisms have been digested by the white blood cell

8. Tissues are groups of the same specialised cells that complete the same function **(1 mark)**; Organs are groups of tissues that are joined together to complete a specific function **(1 mark)**; Systems are groups of organs that work together to complete a specific function **(1 mark)**

9. **(a)** It is responsible for transporting carbon dioxide, oxygen, nutrients, waste products and hormones around the body
   (b) Without oxygen cells cannot respire and produce energy

10. **(a and b)**

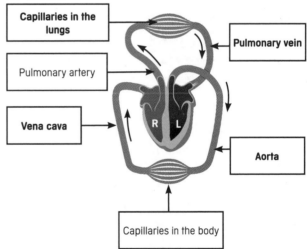

Capillaries in the lungs

Pulmonary vein

Pulmonary artery

Vena cava

Aorta

Capillaries in the body

   (c) The blood passes through the heart twice on a journey around the body **(1 mark)**; It goes from the heart to the lungs, to the heart and to the rest of the body before returning to the heart **(1 mark)**

11. **(a)** Arteries
    (b) Lungs
    (c) Pulmonary vein

12.

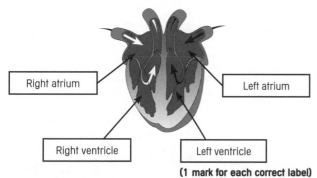

Right atrium
Left atrium
Right ventricle
Left ventricle

(1 mark for each correct label)

13. (a) Blood would not be pumped as effectively to the body
(b) Coronary heart disease (or a heart attack)
(c) Death

14. (a) Because ventricles need to pump the blood to the lungs and rest of body **(1 mark)**; Atria just need to pump blood to the ventricles **(1 mark)**
(b) Because the left ventricle needs to pump blood further (to the body) than the right ventricle (to the lungs)

15. In this order: D, A, C

16. (a) A is an artery; B is a vein; C is a capillary
(b) To cope with blood under high pressure as it has just been pumped from the heart
(c) (i) A vein
    (ii) Arrow pointing upwards **(1 mark)**; Cross above diagram **(1 mark)**
    (iii) To stop blood flowing backwards
(d) (i) To take blood into tissues for exchange of substances
    (ii) (1 mark for structure, 1 mark for function) Their narrow, thin walls allow substances to diffuse easily
(e) Arteries
(f) Arteries
(g) Arteries

17. (a), (b) and **(c)**

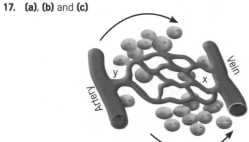

Vein
Artery
y
x

18. (a)

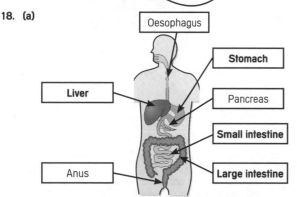

Oesophagus
Stomach
Liver
Pancreas
Small intestine
Anus
Large intestine

(1 mark for each correct label)
(b) To break down large insoluble molecules (food) into smaller ones which can be absorbed into the blood
(c) (i)–(ii) In either order: Provides the correct pH for protease enzymes to work; Kills bacteria
(d) (i) Large intestine
    (ii) Because the large intestine absorbs water, so diarrhoea results from a problem with this organ
(e) To produce bile which helps break down fats

19. In any order:

Enzyme: Carbohydrase   Reactants and products: Breaks down carbohydrates into sugars **(1 mark)**
Enzyme: Lipase   Reactants and products: Breaks down fats into fatty acids and glycerol **(1 mark)**
Enzyme: Protease   Reactants and products: Breaks down proteins into amino acids **(1 mark)**

20. (a) The carbohydrase enzyme (amylase) in her saliva broke the carbohydrate in the bread down into sugars
(b) Saliva in the mouth only contains carbohydrase and not lipase to break down fats (butter)
(c) In the small intestine
(d) (i) Protein is broken down into amino acids **(1 mark)** by proteases **(1 mark)**
    (ii) Carbohydrates are broken down into sugars **(1 mark)** by carbohydrases **(1 mark)**
    (iii) Fats are broken down into fatty acids and glycerol **(1 mark)** by lipases **(1 mark)**

21. (a) Because food is made of large, insoluble molecules which need to be broken down into smaller, soluble ones before they can be absorbed
*(b) Fill a short length of Visking tubing with starch and place in a beaker of water. Test the surrounding water solution with iodine to prove that starch cannot pass through. Fill a second short length of Visking tubing with starch and add some amylase enzyme before placing in a beaker of water. (The amylase should break down the starch into glucose.) Test the surrounding water solution again with iodine to prove that starch cannot pass through. Test the water for glucose using Benedict's solution to prove that glucose can pass through.

22. (a) Non-digestible, functional foods
(b) They stimulate the growth of useful bacteria in the intestines
(c) Accept any suitable answer, e.g. Oligosaccharides

23. The pentadactyl limb is a pattern of limb bones found in all classes of vertebrates with four legs (tetrapods) **(1 mark)**; It includes one proximal bone, two distal bones, several carpals, five metacarpals, and many phalanges **(1 mark)**; Animals have evolved different arrangements of these bones **(1 mark)**; The similarity of this arrangement in all tetrapods indicates that they have evolved from a single ancestor **(1 mark)**

24. (a) It stores the bile made by the liver
(b) Bile neutralises stomach acid **(1 mark)**; Bile emulsifies fat **(1 mark)**

*25. Villi are small finger-like projections that cover the walls of the small intestine. They massively increase the surface area of the intestine. They have a huge blood supply. They are only one cell thick. This allows efficient absorption of small, soluble molecules into the blood.

B3 Control Systems (Pages 68–80)

1. (a)–(b) Accept any two from: Carbon dioxide; Water; Urea

2.

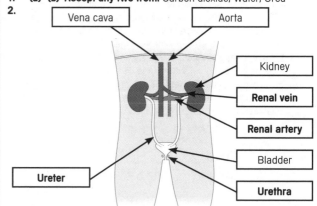

Vena cava
Aorta
Kidney
Renal vein
Renal artery
Bladder
Ureter
Urethra

3. (a) (i)–(ii) In either order: Blood vessels; Tubules
(b) Ureter
(c) Bladder
(d) None
(e) (i) Accept any three from: Water; Ions; Urea; Sugar
    (ii) Filtration
(f) Selective reabsorption

(g) They are passed to the bladder and eventually released from the body as urine.

4. (a) Accept any three from: Vomiting; Diarrhoea; Nausea; Too concentrated and infrequent urine; Too dilute and too much urine

(b) Kidney transplants: the insertion of a donated kidney **(2 marks)**
Kidney dialysis: where small amounts of blood are systematically removed and filtered before being placed back in the body **(2 marks)**

5. (a)

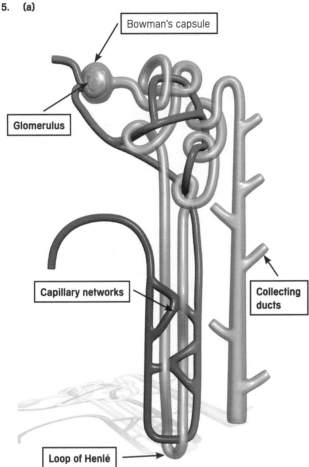

Bowman's capsule

Glomerulus

Capillary networks

Collecting ducts

Loop of Henlé

(b) Filtration: all small molecules and lots of water are squeezed out of the blood and into the tubules
Selective reabsorption: useful substances (water, ions, glucose) are absorbed back into the blood
Excretion of waste: excess water, ions and urea are passed to the bladder before being urinated

6. (a) The kidney does not filter protein from the blood (it is too large to be filtered)

(b) 0% **(1 mark)**; Glucose is absorbed back into the blood during selective reabsorption **(1 mark)**

(c) Water

(d) The useful substances (glucose, ions and water) **(1 mark)** are reabsorbed into the blood **(1 mark)**

(e) Urine

7. **(1 mark for correct thickness of lining, 2 marks for appropriate oestrogen and progesterone concentrations)**

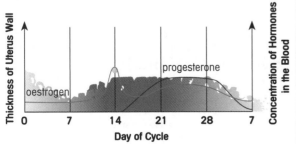

8. (a) Day 0 and day 7

(b) Day 14

(c) In case a fertilised egg implants and grows into a baby

(d) Oestrogen is produced by the ovaries at the beginning of the menstrual cycle **(1 mark)**; It causes the lining of the uterus to thicken ('repairing' itself after a period) **(1 mark)**

(e) (i) The levels of progesterone remain high

(ii) To ensure that a period does not start (which would mean the zygote could not grow into a baby)

9. (a) Sperm cell

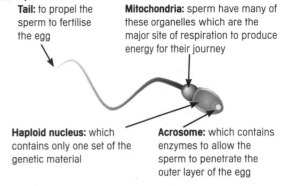

Tail: to propel the sperm to fertilise the egg

Mitochondria: sperm have many of these organelles which are the major site of respiration to produce energy for their journey

Haploid nucleus: which contains only one set of the genetic material

Acrosome: which contains enzymes to allow the sperm to penetrate the outer layer of the egg

(1 mark for each correct label and explanation)

(b) Egg cell

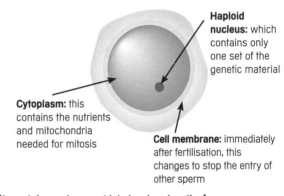

Haploid nucleus: which contains only one set of the genetic material

Cytoplasm: this contains the nutrients and mitochondria needed for mitosis

Cell membrane: immediately after fertilisation, this changes to stop the entry of other sperm

(1 mark for each correct label and explanation)

10. Gametes

*11. Fertile women donate a small number (usually about ten) of their ova anonymously. They are then fertilised *in vitro*. The advantage is low medical risk to the donor. The disadvantage is that it relies on a donor.
In *in vitro* fertilisation, ova are fertilised with sperm outside the female's body before being returned to grow into a baby. The advantage is low medical risk. The disadvantage is high cost.
Surrogate mothers agree to become pregnant for, and deliver a baby to, another couple. The disadvantage is that there are complicated laws surrounding this treatment.

12. (a) Accept any suitable answer, e.g. More expensive to raise two or more babies; Two or more babies are more demanding than one

(b) It would increase

(c) Accept any suitable answer, e.g. Increased demand for resources like food; More housing needed; More jobs needed

*13. All eggs have one X chromosome. Approximately half of all sperm have an X chromosome. The other half have a Y chromosome. If a sperm with an X chromosome fertilises an egg, a female is formed. If a sperm with a Y chromosome fertilises an egg, a male is formed.

14. (a) Milkmaids who contracted cow pox did not normally get small pox

(b) Jenner inoculated (infected on purpose) a small boy's arm with pus from the cow pox virus **(1 mark)**; The boy fell ill but soon recovered **(1 mark)**; Jenner then discovered that the boy could not contract the small pox virus **(1 mark)**

15. (a) The process in which a person's body becomes resistant to infection from a pathogen

(b) Passive immunisation is provided by antibodies from outside the body (from mother's breast milk) **(1 mark)**; Active immunisation results from a body's normal response to infection or as a result of vaccination **(1 mark)**

16. (a) A weakened version of a pathogen (bacteria or virus) is isolated or genetically modified **(1 mark)**; The individual is infected with this (normally in the form of an injection) **(1 mark)**; The person's immune response then begins **(1 mark)**; The body's lymphocytes produce antibodies to destroy the pathogen (often making them clump together) **(1 mark)**

(b) Because her memory lymphocytes recognise the infection **(1 mark)**; They respond by making antibodies far more quickly **(1 mark)**

17. (a) Measles, mumps and rubella

(b) Some scientists thought that there was link between it and autism **(1 mark)**; Recent studies have shown that there is absolutely no link **(1 mark)**

(c) **Accept any suitable answers, e.g.** Advantages: it is an effective way of protecting a large number of people; they are generally inexpensive **(1 mark)**; Disadvantage: a small number of people may fall ill (particularly those with weakened immune systems) **(1 mark)**

*18. Measure two identical samples of milk into two containers. Incubate one at room temperature. Place the other in a fridge. Add a small amount of Resazurin dye to both samples. A quicker colour change from blue to white indicates more bacteria are present. Less might be expected in the sample kept in the fridge.

19. (a) Binary fission

(b) **(1 mark for correct axes, 1 mark for labels, 1 mark for line)**

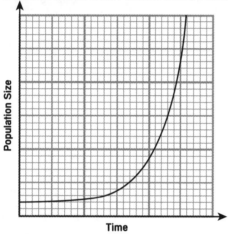

(c) Because growth is limited by a lack of nutrients **(1 mark)** or the production of too many toxic waste products **(1 mark)**

20. (a) Pasteurisation

(b) A method that scientists and medics use to reduce contamination **(1 mark)**; This might involve dipping instruments in alcohol or putting them in a flame **(2 marks)**

(c) Pasteur used sterile broths to prove that microorganisms were responsible for souring wine, beer and milk **(1 mark)**; He then proved that microorganisms could not grow in broths that had been boiled (the basis of pasteurisation) **(1 mark)**; He then proved that microorganisms could not spontaneously generate (as was previously thought) by comparing broths kept sterile after boiling with those open to the air **(1 mark)**

21. (a) **Accept any suitable answer, e.g.** Dutch elm disease caused by a fungus which is spread by a beetle; Chestnut blight caused by a fungus; Tobacco mosaic virus which causes discolouration of leaves of tobacco and similar plants

(b) Some plants produce chemicals to defend themselves from attack.

(c) Plant diseases can affect the food supply **(1 mark)**; **Accept any suitable answer, e.g.** Potato blight in the 19th Century caused mass starvation; Rice blast currently destroys huge amounts of rice **(1 mark)**

22. (a) B **should be ticked.**

(b) Twenty-four hour cycles that some animals and plants live by

23. (a) The pituitary gland

(b) Because the substance that is produced (here ADH) opposes a change to a system (here high salt concentration) **(1 mark)**; This returns the system to normal **(1 mark)**

24. (a) Siona's body determines that her blood salt concentration is too low (her blood water level is too high from the water in the lemonade) **(1 mark)**; The pituitary gland recognises this and releases less ADH into the blood **(1 mark)**; Less water is then reabsorbed into the blood from the tubules **(1 mark)**; Large amounts of dilute urine are released **(1 mark)**

(b) Siona's body determines that her blood salt concentration is too high (her blood water level is too low from sweating) **(1 mark)**; The pituitary gland recognises this and releases more ADH into the blood **(1 mark)**; More water is then reabsorbed into the blood from the tubules **(1 mark)**; Small amounts of concentrated urine are released **(1 mark)**

25. (a) Hormones

(b) D **should be ticked.**

(c) Ovulation would not occur

(d) Low progesterone levels allow FSH to stimulate the maturation of an egg in a follicle **(1 mark)**; This stimulates oestrogen production **(1 mark)**; High levels of oestrogen stimulate a surge in LH which triggers ovulation **(1 mark)**; This triggers the release of progesterone which inhibits FSH and LH production **(1 mark)**

(e) FSH is given to women who are infertile to stimulate the maturation of eggs

26. (a) (i) Memory lymphocytes allow the formation of more antibodies in a shorter time after the second exposure to a pathogen

(ii) **(1 mark for line, 2 marks for labels)**

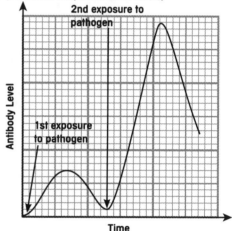

(b) To remind the memory lymphocytes of the antibodies they need to produce

*27. (a) A mouse is vaccinated to stimulate the production of antibodies. Spleen cells are collected from the mouse. Tumour cells (myeloma) are fused with spleen cells to form hybridoma cells. Hybridoma cells are grown in tissue culture. Antibody-forming cells are selected. Monoclonal antibodies are collected.

(b) (i) Monoclonal antibodies are used to identify very small amounts of a hormone called chorionic gonadotrophin **(1 mark)**; This is released by the developing embryo **(1 mark)**; If this hormone is present in a woman's urine, she is pregnant **(1 mark)**

(ii) **Accept any suitable answer, e.g.** They can bind directly to the cancerous cells **(1 mark)**; This encourages the sufferer's immune system to attack them **(1 mark)**

(iii) They are a better way to treat cancer than chemotherapy and radiotherapy **(1 mark)**; as both these treatments have side effects **(1 mark)**; **Accept any two suitable answers, e.g.** lowering the sufferer's immune system, hair loss, tiredness and inability to form blood clots **(2 marks)**

B3 Behaviour (Pages 81–87)

1. A display to attract a member of the opposite sex, for reproduction

2. (a) Only having one mate in a lifetime, rather than a number of different mates

(b) (i)–(ii) Accept any two suitable answers, e.g. Puffins; Swans

3. (a) They care for their young

(b) In any order:

(i) Pregnancy/gestation (offers protection before birth)

     (ii) Breast feeding (provides constant food supply)

     (iii) General protection (to stop predation)

4. **(a) (i)–(ii) Accept any two suitable answers, e.g.** Incubate the eggs; Feed the young; Provide general protection (e.g. building nests)

   (b) Because it increases the chances of the parents' genes being passed successfully to the next generation

5. **(a) (1 mark for labels, 2 marks for points correctly plotted)**

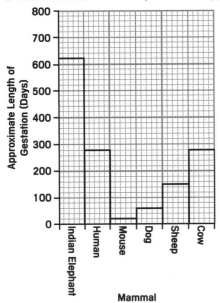

   (b) The larger the animal, the longer the gestation period; (The smaller the animal, the shorter the gestation period)

6. A **should be ticked.**

7. **(a)** Behaviour which is inherited from parents and not affected by the environment

   (b) Accept any suitable answer, e.g. Suckling in mammals

8. **(a)** Classical conditioning is a form of learning without trying; learning by association **(1 mark)**; Operant conditioning is learning as a result of gaining a reward/punishment **(1 mark)**

   (b) Because an animal learns to simply ignore a stimulus that does not affect it

9. Pavlov rang a bell every time he showed a dog some food **(1 mark)**; The dog salivated **(1 mark)**; After a number of occasions, Pavlov rang the bell without showing the dog the food **(1 mark)**; The dog still salivated **(1 mark)**

10. **(a) (1 mark for axes, 1 mark for points, 1 mark for line of best fit)**

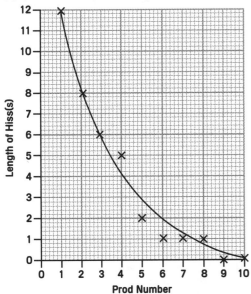

   (b) The length of hiss got shorter as the prod number increased

   (c) Because it had habituated

11. Imprinting occurs in young animals **(1 mark)**; Imprinting is young animals copying their parents **(1 mark)**; **Accept any suitable answer, e.g.** Young birds imprint on their parents immediately after hatching and follow them when they walk **(1 mark)**

12. A choice chamber is a small box which is separated into different areas **(1 mark)**; To investigate behaviour in woodlice, you could place them in the box and see which area they moved to **(1 mark)**; To investigate their preference for damp or dry, you would make one area damp and the other dry **(1 mark)**; You could also do this for light and dark **(1 mark)**

13. **(a) (i)–(ii) Accept any two suitable answers, e.g:** To warn of danger; To work together/cooperation; Courtship

   (b) (i)–(iii) Accept any three suitable answers, e.g: By making sounds; By producing chemicals/pheromones; By giving displays; Through body posture; Through facial expressions

14. **(a) (i)–(ii) Accept any two suitable answers, e.g:** To warn off a predator; To protect territory; To attract a mate

   (b) Communication is still possible even when another animal of the same species is very far away

15. **(a)** Chemicals that an animal releases into the environment for communication with members of the same species

   (b) Accept any suitable answer, e.g. A female dog produces a pheromone after giving birth to give the puppies a feeling of well being and security

16. D **should be ticked.**

17. **(a)** Communication through body posture and position – it may include facial expressions

   (b) Only members of the same species can understand what the body language or facial expressions mean

18. **(a)** (Highly developed) language

   (b) In either order: (i) Spoken **(ii)** Written (usually)

19. **In any order:** Brightly coloured flowers; Strongly scented flowers; By releasing volatile oils when attacked

20. **(a) Nikolaas Tinbergen:** Studied stickleback fish and geese **(1 mark)**; He is best remembered for studying instinctive behaviour of gull chicks **(1 mark)**; He showed they instinctively knew to peck at red spots on their parent's beaks **(1 mark)**

   (b) Konrad Lorenz: Studied geese and jackdaws **(1 mark)**; He is best remembered for studying imprinting **(1 mark)**; This is when young animals copy their parent's behaviour **(1 mark)**

   (c) Dian Fossey: Studied mountain gorillas **(1 mark)**; The gorillas became habituated to her **(1 mark)**; She is also remembered for her work preventing poaching **(1 mark)**

   (d) Jane Goodall: Studies chimpanzees **(1 mark)**; She is best remembered for being the first person to observe that they have distinct personalities **(1 mark)**; She recorded chimpanzees using tools for the first time **(1 mark)**

21. **Accept any three suitable answers, e.g.** Hunted larger animals; Hunted in groups; Lived in larger groups; Did not move so far from their homes

22. **(a)** Analysis of the feet of the female human-like fossilised skeleton called Ardi shows that humans and chimpanzees may have evolved separately from a common ancestor **(1 mark)**; Analysis of the female human-like fossilised skeleton called Lucy shows that it is likely to have walked upright (like humans) but possessed a small skull (like apes) **(1 mark)**

  *(b) Simple tools, such as hand axes, are first thought to have been used about 2.5 million years ago. As humans evolved, the tools used became more complex. Between 10 000 and 6 000 years ago humans were using arrow tips and spear throwers. Between 6 000 and 4 000 years ago over more sophisticated tools were used. This increased sophistication in tools is evidence for evolution.

23. **(a)** When animals or plants evolve together to be dependant upon each other

   (b) Accept any two suitable answers, e.g. Hummingbirds have evolved with the flowers they feed on (the flowers are pollinated by the birds); Snapdragon flowers have evolved to be pollinated by bumblebees (the bees feed on the nectar)

24. **(a)** Mutations in mitochondrial DNA can be used to construct a 'DNA family tree' showing evolution of some species and relationships to other organisms

    (b) 'African Eve' is the common female ancestor that scientists have been able to trace human mitochondrial DNA back to

B3 Biotechnology (Pages 88–95)

1. The alteration of natural biomolecules (molecules produced by living organisms), using science and engineering to provide us with goods and services

2. **(a)** The process by which microorganisms obtain energy from their growth medium and produce other substances through respiration, changing the chemical composition of the medium.

    (b) **(1 mark for reactants, 1 mark for products)**

      Sugar $\xrightarrow{\text{Yeast}}$ Ethanol + Carbon dioxide

    (c) A large vessel used to cultivate microorganisms for the production of biomolecules on a large scale

    (d) **(1 mark for each correct label)**

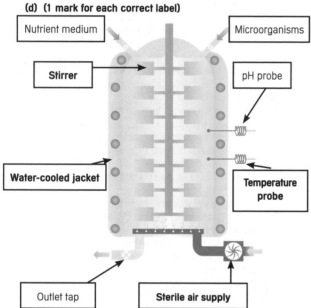

Nutrient medium | Microorganisms | Stirrer | pH probe | Water-cooled jacket | Temperature probe | Outlet tap | Sterile air supply

    (e) **Accept any four from:** Aseptic conditions; Nutrients; Optimum temperature; Correct pH; Oxygenation; Agitation

3. **(a)–(c)** **Accept any three from:** They grow and reproduce quickly; They are easy to handle and manipulate; They can be produced independently of climate; They can make use of waste products from other industrial processes

*4. Place a small amount of yeast into a test tube together with some sugar solution. Incubate for ten minutes and measure the height of the froth that is formed. To make it reliable, repeat the experiment until you get three similar results and take an average. Repeat the whole experiment again at different temperatures, using the same amount of yeast and sugar solution. The temperature at which yeast grows best is that with the highest froth. The independent variable is the temperature. The dependant variable is the height of the froth. The controlled variables are amount of yeast, amount of sugar solution and time.

5. **(a)** Because protein is important for growth and repair

    (b) Because the enzyme chymosin, which was obtained from the stomachs of calves, can now be made by genetically modified microorganisms **(1 mark)**; This means that no animals or animal products are involved in the process, apart from milk **(1 mark)**

6. **(a)** C **should be ticked**.

    (b) Because it contains lactic acid which tastes sour

*7. Heat some milk in a beaker to 20°C. Add 5g of yoghurt starter culture (*Lactobacillus* bacteria) and mix. Cover and incubate. Time how long it takes for the yoghurt to set. Repeat at increasing temperatures (e.g. 30°C, 40°C, 50°C). The controlled variables are: volume of milk, heating time and amount of starter culture.

8. **Accept any suitable answer, e.g.** Add a small quantity of lactase to some milk and refrigerate overnight **(1 mark)**; Keep some of the same milk, to which no lactase has been added, in the same place (to act as the control) **(1 mark)**; Test both samples of milk for the presence of glucose using Benedict's solution **(1 mark)**

*9. Cut an apple into small pieces, weigh it and divide into two equal sections. Add a small amount of pectinase solution to one section, add the same amount of water to the other (to act as the control). Stir both beakers and leave for five minutes. Filter both sections into measuring cylinders and record the volume of juice produced. More juice would be expected in the solution to which pectinase was added.

10. **(a)** Invertase enzymes **(1 mark)**; are used to break down sucrose into glucose and fructose **(1 mark)**

    (b) Protease enzymes are used to break down proteins into amino acids **(1 mark)**; Carbohydrase enzymes are used to break down carbohydrates into sugars **(1 mark)**

11. **In this order:** Scientists find a naturally occurring plant that is resistant to the herbicide **(1 mark)**; Scientists identify the gene that is responsible for the resistance **(1 mark)**; A vector such as *Agrobacterium tumefaciens* is used to transfer the gene for resistance into the genome of the embryo of the crop plant **(1 mark)**; The crop plants are allowed to grow and are resistant to the herbicide **(1 mark)**

12. **(a)** An ethical issue describes an issue on which people have different opinions on whether it is socially and/or morally acceptable

    (b) **Accept any two suitable answers, e.g.** Advantages: increased crop yields; crops can survive with less water **(1 mark)**; Concerns: cross-pollination with native species; the perception that it is 'not natural' **(1 mark)**

    (c) D **should be ticked.**

13. **(a)** Sweet potato: to include a higher amount of Vitamin A **(1 mark)**; To stop blindness in children **(1 mark)**

    (b) Tomato: to contain an antioxidant pigment (flavonoid) **(1 mark)**; This is thought to have anti-cancer properties **(1 mark)**

14. **(a)** Fossil fuels are coal, oil and natural gas **(1 mark)**; Biofuels are made from sustainable resources such as animal waste, organic household waste, wood and alcohol **(1 mark)**

    (b) **Accept any two suitable answers, e.g.** Biofuels are renewable; They are often cheaper than fossil fuels; They still release carbon dioxide when burned but have absorbed it when grown (they are carbon neutral)

    (c) **(1 mark for reactants, 1 mark for products)**

      Ethanol + Oxygen $\longrightarrow$ Carbon dioxide + Water

    (d) **(1 mark for reactants, 1 mark for products)**

      $C_2H_5OH + 3O_2 \longrightarrow 2CO_2 + 3H_2O$

*15. The human gene for insulin is identified. It is removed using restriction enzymes. The same enzyme is used to cut open a section of plasmid DNA. The gene is inserted. The joins are sealed using ligase enzymes. The plasmid is inserted into a bacterium. The bacterium is grown in a fermenter. The bacteria produce human insulin which is collected.

16. **Accept any suitable diagram for 1 mark, e.g.**

    CTCGAGCTAGTGCCGAGCT

    C AGC TCG ATCACGGC TCG A

Restriction enzymes do not cut directly across a double strand of DNA to avoid cutting it into lots of different sections **(1 mark)**; They cut across the double strand of DNA in different places, leaving lengths of single-strand DNA called 'sticky ends' **(1 mark)**; This allows them to be very specific in where they cut DNA **(1 mark)**

17. **In this order:** The gene that is responsible for producing the toxin is identified **(1 mark)**; It is removed using restriction enzymes **(1 mark)**; It is inserted into a vector **(1 mark)**; The vector is inserted into the embryo crop plant's genome **(1 mark)**; The crop plants grow and are resistant to insect pests **(1 mark)**

18. (a) (1 mark for axes, 1 mark for labels, 2 marks for lines of best fit)

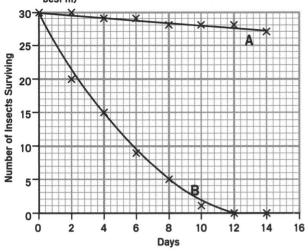

(b) Species B

(c) **Accept any suitable answer, e.g.** Pesticides do not need to be used; Increased levels of food production

(d) **Accept any suitable answer, e.g.** GM is a new and unknown technology; GM crops might spread their genes to other species

2. Describe respiration and ventilation. (2)

....

....

....

3. Steve is running a long-distance race. A graph of his heart rate is shown below.

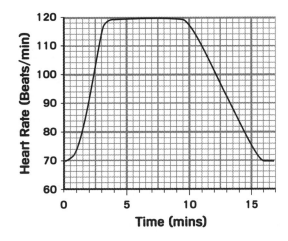

(a) Describe what happens to Steve's heart rate when he starts running the race. (1)

....

(b) State what will happen to Steve's breathing rate. (1)

....

(c) Explain your answers to parts **(a)** and **(b)**, using your knowledge of aerobic respiration. (4)

....

....

....

....

4. Nicole is playing football. She sprints the length of the pitch to score a goal. However, she can barely celebrate because her legs have gone weak and rubbery and she can't get her breath back. Nicole has been respiring anaerobically.

(a) State the word equation for this type of respiration. (2)

....

(b) Explain why Nicole's legs felt rubbery. (2)

....

....

....

5. State which method of respiration is the most efficient. (1)

....

6. **(a)** When does excess post-exercise oxygen consumption occur? (1)

....

(b) Explain why deep breathing occurs during post-exercise oxygen consumption. (2)

....

....

7. **(a)** State **two** ways in which leaves are adapted to make them efficient at photosynthesis. (4)

....

....

....

....

(b) Label the parts of the leaf below. Describe the function of each part. (3)

(c) State the word equation for photosynthesis. (2)

....

(d) State **three** uses of glucose made during photosynthesis. (3)

8. In a biology lesson, William was investigating how various factors affect the rate of photosynthesis. He looked at temperature first. The results he obtained are shown below.

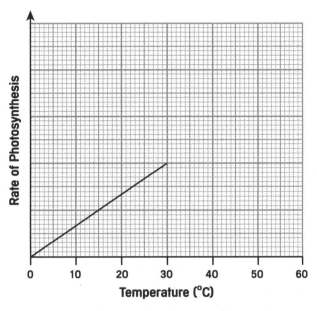

(a) He did not finish drawing the graph. Complete the graph to show the results he might have expected. (1)

(b) He then looked at how the rate of photosynthesis changed on two different days.

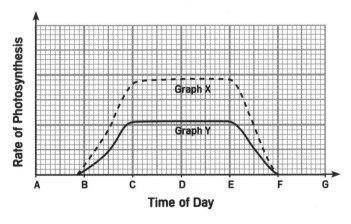

(i) State which line, X or Y, represents a warm, sunny day. (1)

(ii) Label the graph to show where dawn and nightfall occur. (2)

(iii) Describe in detail the similarities in Graph X and Graph Y in **(b)** and explain the differences between them. (4)

....

....

....

....

....

(c) (i) William added extra carbon dioxide to aquatic plants kept in his fish tank. He counted the bubbles of oxygen produced to determine the rate of photosynthesis.

Draw a graph to show the effects of increasing carbon dioxide on the rate of photosynthesis. Label both axes. (3)

(ii) Describe in detail what the graph in **(c) (i)** shows. (3)

....

....

....

....

9. **(a)** Explain how diffusion is different from osmosis. (2)

....

....

....

(b) The diagram below shows an experiment to demonstrate osmosis. Explain the results in detail.   (3)

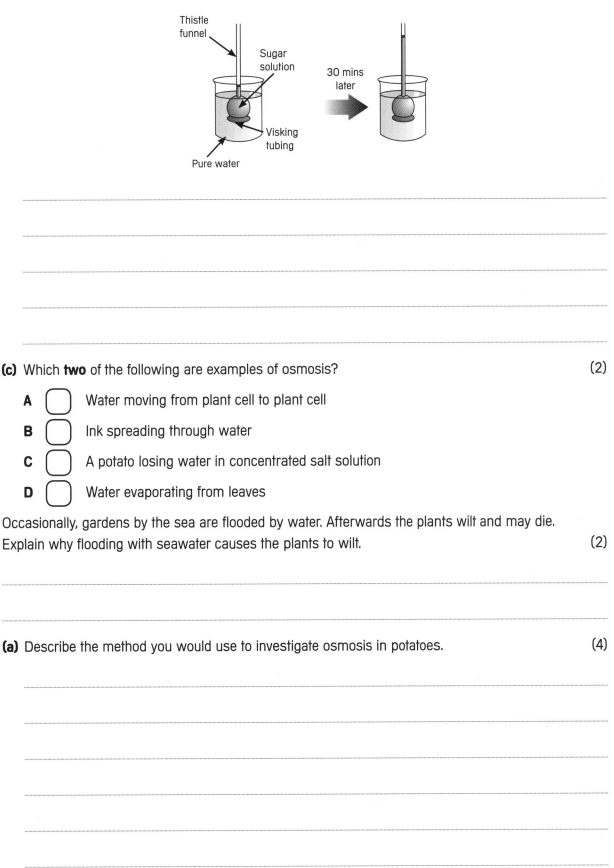

Thistle funnel

Sugar solution

30 mins later

Visking tubing

Pure water

....

....

....

....

....

(c) Which **two** of the following are examples of osmosis?   (2)

A ⬭ Water moving from plant cell to plant cell

B ⬭ Ink spreading through water

C ⬭ A potato losing water in concentrated salt solution

D ⬭ Water evaporating from leaves

10. Occasionally, gardens by the sea are flooded by water. Afterwards the plants wilt and may die. Explain why flooding with seawater causes the plants to wilt.   (2)

....

....

11. **(a)** Describe the method you would use to investigate osmosis in potatoes.   (4)

....

....

....

....

....

....

(b) Describe the results you would expect to find. (2)

_____

_____

12. **(a)** State the definition of the term **biodiversity**. (1)

_____

(b) Explain why scientists sample populations. (1)

_____

_____

(c) State the **two** groups that methods of sampling can be divided into. (2)

_____

13. Nic and Debbie's gardens back on to each other. Nic's lawn is shaded by his house, whilst Debbie's is not. They decided to investigate the differences in plant species that grew in their lawns.

(a) Describe how they would do this. Name any equipment they would use in your answer. (3)

_____

_____

_____

_____

(b) State what effect using a greater number of quadrats will have on their data. (1)

_____

(c) Explain why they cannot definitely say that shade is responsible for the differences in the plants that have grown in their lawns. (3)

_____

_____

_____

14. (a) State the major way in which plants lose water. (1)

....

(b) Describe what a plant can do to reduce water loss. (1)

....

(c) (i) State what process will be slowed down as a result of water loss. (1)

....

(ii) Explain why. (1)

....

*15. Describe water transport in plants. (6)

....

....

....

....

....

....

....

....

....

16. The diagram below shows an experiment investigating the water uptake by a plant. The results table is also shown below.

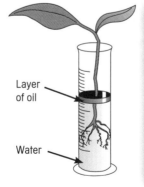

Layer of oil

Water

Time from Start (Days)	Volume of Water in Cylinder (cm³)
0	50
1	47
2	43
3	42
4	40

(a) Why is a layer of oil placed in the measuring cylinder? (1)

....

....

(b) Explain why the volume of water in the measuring cylinder has reduced. (1)

_____

_____

(c) State how you would expect the results to have changed if:

   (i) the plant had been in a colder room (1)

_____

   (ii) the plant had been in a more humid atmosphere (1)

_____

   (iii) air had been blown over the leaves of the plant (1)

_____

   (iv) the plant had been placed in the sunshine. (1)

_____

17. There are a number of other sampling methods that do not use quadrats.

(a) Describe each of the sampling methods below.

   (i) Pitfall trap (2)

_____

_____

_____

   (ii) Sweep netting (2)

_____

_____

_____

   (iii) Kick sampling (2)

_____

_____

_____

(b) (i) Draw and label a diagram of a pooter in the space below. (2)

(ii) Describe how it works. (1)

....

....

....

18. Describe how scientists use the mark, release, recapture method to sample mobile animals. (3)

....

....

....

....

19. Scientists closely monitor the environment.

(a) State **one** factor they might measure. (1)

....

(b) Describe why they might use electronic equipment to do this. (1)

....

(Total: .... / 110)

Questions labelled with an asterisk (*) are ones where the quality of your written communication will be assessed – you should take particular care with your spelling, punctuation and grammar, as well as the clarity of expression, on these questions.

1. **(a)** State what **fossils** are. (1)

    (b) State **two** reasons why the fossil record is incomplete. (2)

2. Describe the processes listed below.

    (a) Cell specialisation (1)

    (b) Cell elongation (1)

    (c) Cell division (1)

3. **(a)** State why length and height are not accurate measurements of growth. (1)

    (b) Explain why wet mass is usually used to measure growth, when dry mass is a more accurate measurement. (1)

4. Draw graphs to show what you would expect the rate of growth for **(a)** an elephant, and **(b)** a redwood tree, to be over their lifetime.

(a) 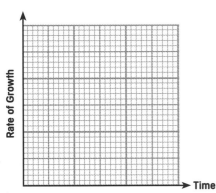   **(b)** 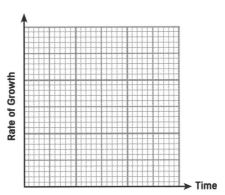   (2)

(c) Describe the difference (if any) in the shapes of your graphs.   (2)

....

....

....

5. State **four** components of blood.

(a) ....   (1)

(b) ....   (1)

(c) ....   (1)

(d) ....   (1)

6. Blood carries oxygen around the body.

(a) State the part of the blood that carries the oxygen.   (1)

....

(b) Describe **two** features of this part of the blood that make it well adapted to this function.   (4)

....

....

(c) Describe how blood cells carry the oxygen from the lungs to the organs in the body.   (2)

....

....

....

....

7. Describe, using diagrams to illustrate your answer, how phagocytes function. (3)

8. State the definitions of the terms **tissues**, **organs** and **systems**. (3)

9. **(a)** State the function of the circulatory system. (1)

(b) A patient has lost a lot of blood. Unless she gets more blood, her cells will die. Explain why. (1)

10. **(a)** Using the word list below label the diagram of the circulatory system. (4)

| Vena cava | Aorta | Capillaries in the lungs | Pulmonary vein |

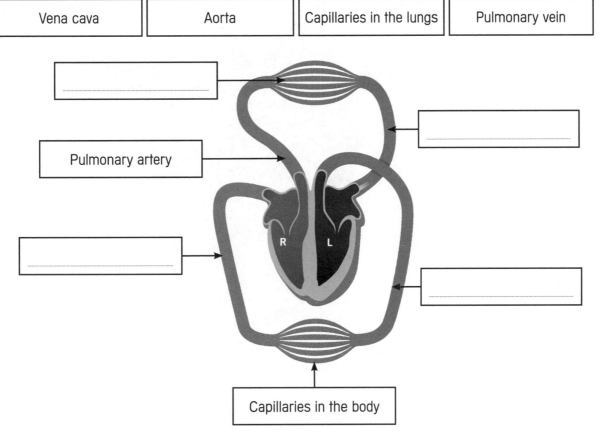

Pulmonary artery

R L

Capillaries in the body

(b) Clearly label, with arrows, the direction of blood flow around the circulatory system on the diagram above. (4)

(c) This system is called a double circulatory system. Describe what this means. (2)

11. State the following parts of the body.

(a) The vessels that carry blood away from the heart .... (1)

(b) The place where gaseous exchange occurs .... (1)

(c) The only vein to carry oxygenated blood .... (1)

12. Label the diagram of the heart below. (4)

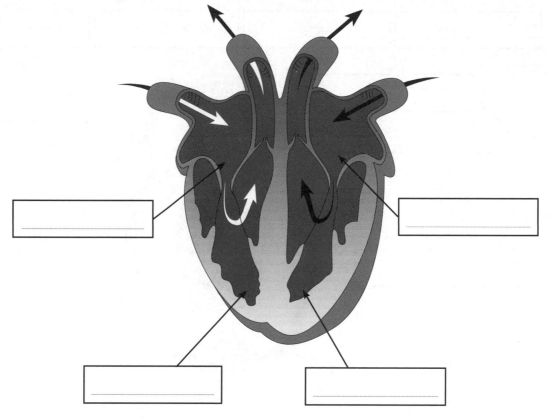

13. (a) Describe what would happen if the valve in the left side of the heart became damaged. (1)

....

(b) State what medical condition might occur if one of the blood vessels supplying the heart's muscular walls became narrowed. (1)

....

(c) State the ultimate effect that either of the above would have on a person. (1)

....

....

14. (a) Explain why the ventricles have larger and more muscular walls than the atria. (2)

....

....

(b) Explain why the left ventricle has a more muscular wall than the right ventricle. (1)

....

....

15. Rearrange the following statements into the right order so that they explain how blood is pumped by the heart. (3)

A The ventricles are relaxed and rapidly fill with blood.

B The atria relax allowing blood to return to the heart from the vena cava and the pulmonary vein.

C The ventricles contract, forcing blood out of the heart to the lungs and body.

D The atria contract forcing blood into the ventricles through the two heart valves.

B → ☐ → ☐ → ☐

16. The diagram below shows the three types of blood vessel.

A   B   C

(a) State their names. (3)

A .... B .... C ....

(b) State why blood vessel **A** has a thick elastic, muscular wall. (1)

....

(c) The diagram below shows a valve in a blood vessel.

   (i) State what type of blood vessel this is. .... (1)

   (ii) Draw an arrow on the diagram to show the direction of flow of the blood. Mark with an X the direction of the heart. (2)

(iii) State the purpose of the valve. .... (1)

(d) (i) State the function of blood vessel **C**. (1)

....

   (ii) Describe how its structure makes it well adapted for this purpose. (2)

....

....

(e) State the name of the blood vessels that take blood away from the heart. (1)

....

(f) State the name of the blood vessels in which we can feel a pulse. (1)

_____

(g) State the name of the blood vessels that carry blood at high pressure. (1)

_____

17. The diagram below shows a capillary network in a muscle.

(a) Draw arrows on the diagram to show (1)
the direction of blood flow through
the muscle.

(b) Mark with an X where the blood would be (1)
rich in carbon dioxide and waste.

(c) Mark a Y where the blood would be rich in oxygen (1)
and food.

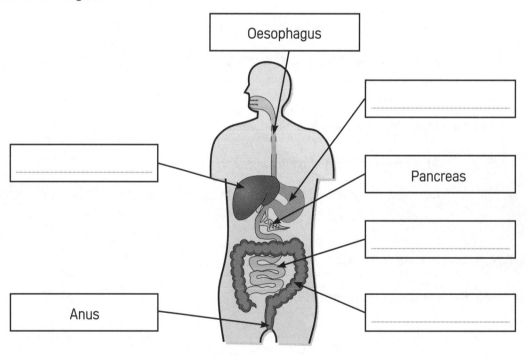

18. The diagram below shows some parts of the digestive system.

(a) Label the diagram. (4)

Oesophagus

Pancreas

Anus

(b) State the function of the digestive system. (1)

_____

(c) The stomach contains an acid. State **two** of the functions of acid in the stomach.

(i) _____ (1)

(ii) _____ (1)

(d) (i) George has diarrhoea. The doctor says it is due to an infection of the digestive system. State which part of George's digestive system has been affected. (1)

(ii) Explain your answer. (1)

(e) State the function of the liver in the digestive system. (1)

19. State the reactants and products that the three enzymes catalyse in the digestive system. (3)

Enzyme: _____Carbohydrase_____   Reactants and products: _____

Enzyme: _____Lipase_____   Reactants and products: _____

Enzyme: _____Protease_____   Reactants and products: _____

20. Emily goes to a restaurant for lunch. She eats a bread roll with butter.

(a) Explain why the bread started to taste sweet after she had chewed it for a long time. (1)

(b) Explain why the digestion of butter did not start in Emily's mouth. (1)

(c) State where the digestion of butter would have started. (1)

(d) Emily also ate a hamburger in a bun and drank a milkshake. This meal is a source of protein, carbohydrate and fat. State what enzyme acts upon these substances and what they are broken down into.

(i) Protein (2)

(ii) Carbohydrate (2)

(iii) Fat (2)

21. **(a)** Explain why food molecules must be digested before they can enter the blood. (1)

*(b) Describe an experiment involving enzymes in which Visking tubing is used to model the digestive system. (6)

22. **(a)** State what **prebiotics** are. (1)

(b) What effect do they have on the digestive system? (1)

(c) Name **one** prebiotic. (1)

(Total: .... / 106)

23. Explain how the pentadactyl limb is used as evidence for evolution. (4)

24. (a) Some people suffer from gall stones. This condition can require the removal of the gall bladder. State the function of the gall bladder. (1)

(b) State the **two** functions of bile. (2)

*25. Villi line the small intestine. Explain how their structure aids their function. (6)

(Total: .... / 13)

Questions labelled with an asterisk (*) are ones where the quality of your written communication will be assessed – you should take particular care with your spelling, punctuation and grammar, as well as the clarity of expression, on these questions.

1. Cells produce a number of waste products. State **two** of them. (2)

(a) _____ **(b)** _____

2. Label the diagram below. (4)

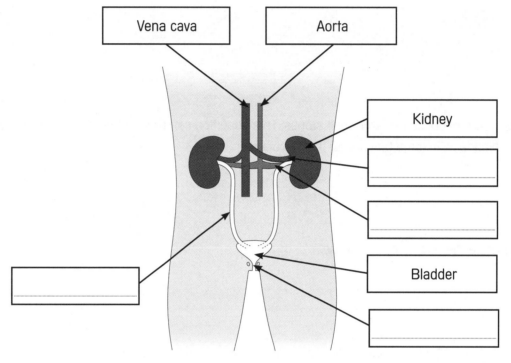

| Vena cava | Aorta |

| Kidney |

| Bladder |

3. The kidney helps to maintain the internal environment of the body.

(a) State the **two** important tissues in the kidney. (2)

(i) _____ **(ii)** _____

(b) State the name of the tube that takes urine from the kidney. (1)

_____

(c) State the name of the organ where urine is stored. _____ (1)

(d) State the amount of glucose you would normally expect to find in urine. _____ (1)

(e) (i) State **three** substances that pass out of the blood into the tubules. (3)

_____

(ii) State the name given to this process. _____

(f) Some substances that the body cannot afford to lose are taken back into the blood.

State the name given to this process. _____ (1)

(g) Describe what happens to all the excess water, ions and urea. (1)

....

4. **(a)** State **three** symptoms of kidney failure. (3)

....

....

(b) Describe the **two** treatments that exist for kidney failure. (4)

....

....

....

....

5. **(a)** Label the diagram below of a nephron. (4)

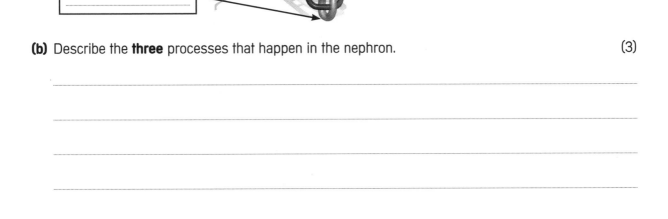

Bowman's capsule

(b) Describe the **three** processes that happen in the nephron. (3)

....

....

....

....

....

6. The table below shows the percentage concentration of various substances in the blood plasma and in the liquid filtered out of the blood into the Bowman's capsule.

Substance	Blood Plasma %	Liquid Filtered out of the Blood into the Bowman's Capsule (%)
Urea	0.03	0.03
Ions	0.4	0.4
Glucose	0.1	0.1
Protein	8.0	0

(a) Explain why no protein is found in the liquid filtered by the kidneys. (1)

(b) State what percentage concentration of glucose you would expect to find in the liquid stored in the bladder. Explain your answer. (2)

(c) State the other important substance missing from the table. (1)

(d) State the definition of the term **selective reabsorption**. (2)

(e) State the name of the liquid stored in the bladder. (1)

7. Complete the graph below of the menstrual cycle. (3)

You must include:

- a timescale

- thickness of the uterus wall

- concentration of hormones in the blood.

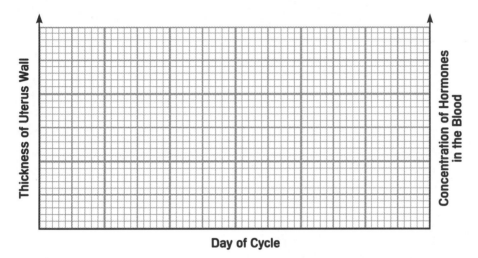

Day of Cycle

8. **(a)** Between which days of the menstrual cycle does the uterus wall break down? (1)

....

(b) On which day is the egg released? (1)

....

(c) Explain why the uterus wall becomes lined with blood vessels. (1)

....

(d) Describe the changes that occur to the levels of the hormone oestrogen during the menstrual cycle. Link the changes to the thickness of the uterus lining. (2)

....

....

....

(e) **(i)** State what happens to the levels of progesterone if a fertilised egg becomes implanted in the wall of the uterus. (1)

....

(ii) Explain why this happens. (1)

....

9. Label the diagrams of a sperm and egg cell below. Explain how each cell is adapted for its function.

(a) (4)

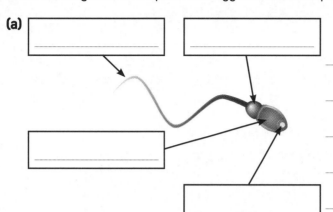

(b) (3)

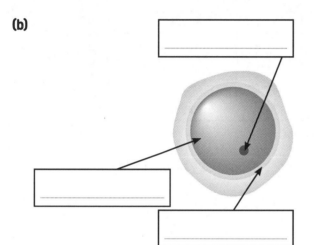

10. State the collective name for sperm and egg cells. (1)

*11. State **three** methods of treating infertility in humans. Describe an advantage or disadvantage of each treatment method in your answer. (6)

12. Below is some information about IVF treatment.

> In normal pregnancies, no more than 2% of women have a multiple birth, i.e. give birth to twins, triplets, etc. With IVF treatment, the chance of having a multiple birth rises to 25%.

(a) Describe **one** problem that a multiple birth might cause for a couple. (1)

....

....

(b) If IVF treatment becomes more common, the number of multiple births will increase. State **one** effect that this would have on the population. (1)

....

(c) Suggest **one** way in which this could impact on society. (1)

....

....

*13. Describe how the gender of a baby is determined by the fertilising sperm. (6)

....

....

....

....

....

....

....

....

....

....

....

....

14. (a) State the observation that Jenner made in the 18th Century before his discovery of the small pox vaccine. (1)

(b) Describe how Jenner proved his discovery. (3)

15. (a) State the definition of the term **immunisation**. (1)

(b) Describe the difference between active and passive immunisation. (2)

16. (a) Melissa visits the doctor to have her annual flu vaccination. She does not understand the doctor's explanation of a vaccination. Explain to Melissa how a vaccination protects her from infection. (4)

(b) Melissa will never catch the same flu virus twice. Explain why. (2)

17. Recently there has been controversy surrounding the MMR vaccine.

(a) State the diseases that this vaccination protects children against. (1)

(b) Explain why the vaccination was once controversial but is less so now. (2)

(c) State **one** advantage and **one** disadvantage of vaccinations. (2)

*18. You want to investigate whether keeping milk in a fridge reduces the growth of bacteria. Describe an experiment you could conduct to investigate this. Describe what results you would find. (6)

19. **(a)** What is the name of the way in which bacteria multiply? (1)

(b) Draw a graph to show the changes in the size of a bacterial population over time, provided it has appropriate conditions and nutrients. Label both axes. (3)

(c) Explain why bacteria do not grow like this for ever. (2)

_____

_____

_____

20. Louis Pasteur was a French scientist who lived in the 19th Century.

(a) Besides aseptic technique, what else is Pasteur remembered for? (1)

_____

(b) Describe what **aseptic technique** is. State **two** examples of it in your answer. (3)

_____

_____

_____

(c) Describe the experiments in which Pasteur used sterile broths. (3)

_____

_____

_____

_____

_____

21. **(a)** State a plant disease caused by a pathogen. Give both the plant and the pathogen in your answer. (2)

_____

_____

_____

(b) Describe how plants can defend themselves from attack by pathogens. (1)

_____

_____

_____

(c) State **one** economic effect of plant diseases. Give an example in your answer. (2)

22. **(a)** Most plants are able to detect changes in the length of daylight. This is called: (1)

A ◯ heterotrophic B ◯ photoperiodicity

C ◯ poikilothermic D ◯ photosynthesis.

(b) State the definition of the term **circadian rhythms**. (1)

(Total: .... **/ 114)**

Higher Tier

23. **(a)** State where the hormone ADH is produced. .... (1)

(b) Explain why the role of ADH in the kidney is an example of negative feedback. (2)

24. **(a)** Siona drinks two pints of lemonade over a period of 30 minutes. Describe how her body responds to this intake of fluid. (4)

continued...

(b) The next day is hot and sunny. Siona plays tennis for two hours without having a drink. Describe how her body responds to the steady loss of fluid through sweating. (4)

25. **(a)** State what type of substances FSH and LH are. (1)

(b) What would happen if a woman was unable to produce FSH? (1)

A ◯ She would have longer periods.

B ◯ Ovulation would not occur.

C ◯ She would have twins.

D ◯ Her eggs would not mature in a follicle.

(c) State what would happen if a woman was unable to produce LH. (1)

(d) Explain the role of FSH and LH in the menstrual cycle. (4)

continued...

(e) Describe how FSH is used to treat infertile women. (1)

26. **(a) (i)** Describe the production of antibodies in response to two subsequent exposures to the same pathogen. (1)

(ii) Draw a graph to show the antibody levels in response to these two exposures. Indicate on the graph where the exposures are. (3)

Time

Antibody Levels

(b) Explain why boosters are sometimes needed after vaccinations. (1)

continued...

*27. (a) Describe how monoclonal antibodies are made. (6)

(b) (i) Describe how monoclonal antibodies can be used to diagnose pregnancies. (3)

(ii) Describe how monoclonal antibodies can be used to treat cancer. (2)

(iii) Describe the advantages of using monoclonal antibodies over other methods for treating cancer. Name the other methods in your answer. (4)

(Total: .... / 39)

Questions labelled with an asterisk (*) are ones where the quality of your written communication will be assessed – you should take particular care with your spelling, punctuation and grammar, as well as the clarity of expression, on these questions.

1. State what **courtship behaviour** is. (1)

    ....

    ....

2. **(a)** State what **monogamy** means. (1)

    ....

    ....

    (b) Give **two** examples of monogamous animals. (2)

    (i) ....

    (ii) ....

3. **(a)** Mammals show parental care. State what this means. (1)

    ....

    (b) Describe **three** things that mammalian parents do as part of parental care. (3)

    (i) ....

    (ii) ....

    (iii) ....

4. **(a)** Describe **two** things that birds do to look after their young.

    (i) .... (1)

    (ii) .... (1)

    (b) Explain why parental care has evolved. (1)

    ....

    ....

    ....

    ....

5. Look at the data in the table below.

(a) Draw a bar chart to represent the data in the table. (3)

Mammal	Approximate Length of Gestation (Days)
Indian elephant	620
Human	280
Mouse	20
Dog	60
Sheep	150
Cow	280

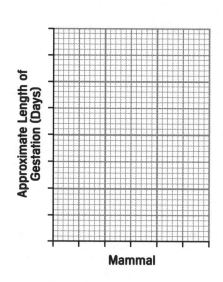

(b) Describe the relationship between the size of the mammal and the gestation period. (1)

....

....

6. Tick the statement that is false. (1)

A ◯ Monogamy is common among mammals.

B ◯ Most animals have a number of different mates.

C ◯ An alpha male mates with all the females in his pride or harem.

D ◯ Animals are born with instinctive behaviours.

7. **(a)** Describe what instincts are. (1)

....

....

(b) State an example of an instinctive behaviour. (1)

....

8. **(a)** Describe classical conditioning and operant conditioning. (2)

....

....

....

....

(b) Explain why habituation is thought to be the simplest form of learning. (1)

....

....

9. Describe the experiment that Pavlov conducted to illustrate classical conditioning. (4)

....

....

....

....

10. Patti discovered that if she prodded a giant cockroach with her finger, it made a hissing sound. She decided to investigate the effect that prodding had on how long the cockroach hissed. She prodded the cockroach every 20 seconds and timed how long it hissed for. Her results are shown below.

Prod Number	1	2	3	4	5	6	7	8	9	10
Length of Hiss (Seconds)	12	8	6	5	2	1	1	1	0	0

(a) Draw a graph to show the results. (3)

(b) Describe what these results show. (1)

....

....

(c) Explain why the cockroach did not hiss after the eighth prod. (1)

....

....

11. Describe when imprinting behaviour occurs and what it is. Give an example of animals that imprint in your answer. (3)

....

....

....

12. Describe what a choice chamber is and how you would use it to investigate **two** different ways in which woodlice behave. (4)

....

....

....

....

....

....

13. **(a)** Give **two** reasons why animals communicate.

(i) .... (1)

(ii) .... (1)

(b) State **three** ways in which animals can communicate.

(i) .... (1)

(ii) .... (1)

(iii) .... (1)

14. **(a)** Give **two** reasons why an animal would make a sound.

(i) .... (1)

(ii) .... (1)

(b) State the advantage to an animal of being able to communicate over long distances. (1)

....

....

15. **(a)** State what **pheromones** are. (1)

....

....

(b) State a specific reason why an animal would release a pheromone. (1)

....

16. Male rabbits and hares thump their hind legs on the ground to: (1)

A ◯ attract a mate

B ◯ show submission

C ◯ threaten other males

D ◯ warn other nearby animals.

17. Describe each of the following:

(a) Body language (1)

....

....

(b) Species-specific communication (1)

....

....

18. **(a)** Humans have the most complex system of communication of all animals. State the special feature that makes it the most complex. (1)

....

(b) State **two** examples of how this special feature is used.

(i) .... **(ii)** .... (2)

19. State **three** ways in which plants communicate with animals. (3)

....

....

....

20. Describe the work of the four famous ethologists named below. For each one, name the animals they studied and what they are best remembered for.

(a) Nikolaas Tinbergen (3)

....

....

....

(b) Konrad Lorenz (3)

....

....

....

(c) Dian Fossey (3)

....

....

....

(d) Jane Goodall (3)

....

....

....

21. Describe the effect that the Ice Ages had on human behaviour. (3)

....

....

....

22. Emma goes to a museum to learn about human history. Whilst there, she sees an exhibit about fossils and early tools.

(a) Describe how fossils Ardi and Lucy provide evidence for human evolution. (2)

....

....

....

....

*(b) Describe how tools provide evidence for human evolution. (6)

(Total: .... / 79)

23. (a) State the definition of the term **co-evolution**. (1)

(b) State **two** examples of co-evolution. For each example, explain why this has occurred. (4)

24. (a) Describe how mitochondrial DNA helps in the study of evolution. (1)

(b) What is **African Eve**? (1)

(Total: .... / 7)

Questions labelled with an asterisk (*) are ones where the quality of your written communication will be assessed – you should take particular care with your spelling, punctuation and grammar, as well as the clarity of expression, on the questions.

1. State the definition of the term **biotechnology**. (1)

....

....

2. **(a)** State the definition of the term **fermentation**. (1)

....

....

....

(b) State the word equation for fermentation. (2)

....

(c) Describe what a fermenter is. (2)

....

....

(d) Label the parts of the fermenter shown below. (4)

| Nutrient medium | Microorganisms |

| | |
| .... | |

| pH probe |

| | |
| .... | .... |

| Outlet tap |

| |
| .... |

(e) State **four** factors or conditions that fermenters require. (4)

....

....

....

....

3. State **three** advantages of using microorganisms in food production.

(a) .... (1)

....

(b) .... (1)

....

(c) .... (1)

....

*4. Describe an experiment in which you investigate the effect of temperature on the growth of yeast. State how you will ensure reliable results. Identify the dependant, independent and controlled variables. (6)

....

....

....

....

....

....

....

....

....

....

....

5. Mohamed is a vegetarian.

(a) He eats a lot of mycoprotein in his diet. Explain why he needs to eat protein. (1)

(b) He is careful to only eat vegetarian cheese. Explain why vegetarians can eat this. (2)

6. **(a)** What process is responsible for turning milk into yoghurt? (1)

A ◯ Genetic engineering

B ◯ Reproduction

C ◯ Fermentation

D ◯ Photosynthesis

(b) Explain why yoghurt tastes sour. (1)

*7. Describe an experiment in which you investigate the effects of temperature on yoghurt making. Mention variables you have controlled in your experiment. (6)

8. Describe an experiment in which you investigate the effect of lactase on milk. Mention a control in your experiment. (3)

*9. Describe an experiment in which you investigate the effect of pectinase on fruit juice. Mention a control in your experiment. Describe the result you would expect. (6)

10. Enzymes are referred to as biological catalysts. Describe how they are used in:

(a) Sweet manufacture (2)

(b) Washing powders (2)

11. Scientists are able to breed herbicide-resistant crops using the vector *Agrobacterium tumefaciens*. Describe this process. (4)

12. The use of genetically modified (GM) crops is an ethical issue.

(a) Describe what is meant by the term **ethical issue**. (1)

(b) State **one** advantage of and **one** concern with genetic modification. (2)

(c) Which of the following is an advantage of GM crops? (1)

A ⬭ Using herbicide-resistant crops could result in 'super-weeds' developing.

B ⬭ We do not know what the long-term health effects are of eating GM crops.

C ⬭ GM crops could affect the biodiversity of habitats.

D ⬭ Some GM crops can survive in poor soils.

13. Genetic modification has been used to alter a number of plants. For both plants below, describe the way in which they have been genetically modified, and give reasons why.

(a) Sweet potato (2)

(b) Tomato (2)

_____

_____

_____

14. More people are starting to use biofuels in place of fossil fuels.

(a) State what **biofuels** and **fossil fuels** are. (2)

_____

_____

_____

_____

_____

(b) Describe **two** advantages of using biofuels. (2)

_____

_____

_____

_____

(c) State the word equation for burning ethanol. (2)

_____

(d) State the balanced symbol equation for burning ethanol. (2)

_____

(Total: _____ / 67)

*15. Recombinant DNA technology is used in the production of insulin for diabetics. Describe how this is made. (6)

16. Restriction enzymes are highly specific in how they cut DNA. Describe, using a diagram, how they do this. (4)

17. The bacterium *Bacillus thuringiensis* produces a toxin that kills many insects. Describe how this microorganism has been used in genetic engineering to produce insect-resistant crops. (5)

continued...

18. An insect-resistant GM species of plant was compared to a non-resistant species. Each plant had 30 pest insects placed on it and their numbers were counted every day for 14 days. The numbers of insects left on each plant are shown in the table below.

Day	0	2	4	6	8	10	12	14
Species A	30	30	29	29	28	28	28	27
Species B	30	20	15	9	5	1	0	0

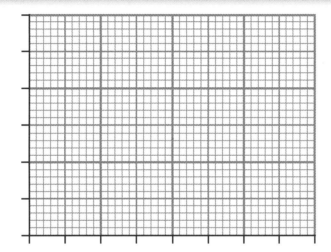

(a) Draw a line graph to represent the data in the table. (4)

(b) State which species is the GM one. (1)

....

(c) Suggest **one** reason why breeding GM insect-resistant plants is a good idea. (1)

....

....

(d) Suggest **one** reason why some people are against producing GM crops. (1)

....

....

(Total: .... / 22)

Notes